Cover Cropping for Vegetable Production

A Grower's Handbook

Technical Editors

Richard Smith
Robert L. Bugg
Mark Gaskell
Oleg Daugovish
Mark Van Horn

University of California
Agriculture and Natural Resources

Publication 3517

To order or obtain ANR publications and other products, visit the ANR Communication Services online catalog at http://anrcatalog.ucdavis.edu or phone 1-800-994-8849. You can also place orders by mail or FAX, or request a printed catalog of our products from

University of California
Agriculture and Natural Resources
Communication Services
1301 S. 46th Street
Building 478 - MC 3580
Richmond, CA 94804-4600
Telephone 1-800-994-8849
510-665-2195
FAX 510-665-3427
E-mail: danrcs@ucdavis.edu

Publication 3517
ISBN-13: 978-1-60107-679-3
Library of Congress Control Number: 2011933957

Photo credits are given in the captions. Cover photos: Jack Kelly Clark, Chuck A. Ingels, Joseph M. DiTomaso, and Richard Smith. Design by Celeste A. Rusconi.

This publication has been anonymously peer reviewed for technical accuracy by University of California scientists and other qualified professionals. This review process was managed by ANR Associate Editor for Vegetable Crops Jeffrey Mitchell.

FSC MIX Paper FSC® C009319

Printed in the United States of America on FSC Certified paper.

2.5m-pr-10/11-SB/CR

Contents

Cover crops can be important resources for the growing of vegetable commodities, and they play critical roles in the ecology and production environment of such crops. *Cover Cropping for Vegetable Production* provides comprehensive information on various aspects of the use of cover crops for vegetable production systems. This manual is intended for growers, pest control advisers, seed company staff, and others who wish to better understand the impacts of cover crops on the overall farm environment, soils, pests, and nutrient cycling, as well as the economics of production. Our intent is to provide research-based information on cover crops and their use. Information on soil and plant interactions is constantly evolving, and this manual presents our current understanding of the role that cover crops play in maintaining crop productivity in intensive vegetable crop production systems.

This publication is a collaborative effort by the authors and many others who provided support and insight. We greatly appreciate the opportunities we had to work with growers, seed companies, and technical agencies to conduct the cover crop research described in this publication. We acknowledge the efforts of the many photographers, associate editor Jeff Mitchell, and ANR Communication Services in bringing this publication to fruition.

Our thanks go to the following individuals for providing information and technical assistance for this publication: John Bauer, Richard Casale, Karen Christensen, Ramy Colfer, Ellen Dean, Joe DiTomaso, Sam Earnshaw, Phil Foster, Steve Hearne, Tom Hearne, Lee Jackson, Tom Johnson, Jim Leap, Joji Muramoto, Phil Radspinner, Allen Steigerwald, Fred Thomas, Bill White, Jonas White, and Ron Yokota.

We also thank the following companies and organizations for their contributions towards the production of this publication: Ag Seeds Unlimited, Woodland, California; Gowan Seeds, Salinas; Kamprath Seed Company, Manteca; L.A. Hearne Company, King City; Peaceful Valley Farm Supply, Grass Valley; Resource Seed Inc., Gilroy; Snow Seed Company, Salinas; Timothy Stewart and Lekos Seeds, Woodland; and White Seed Company, Oxnard.

Contributing Authors

Brenna J. Aegerter, University of California Cooperative Extension Vegetable Crops Advisor, San Joaquin County
José L. Aguiar, University of California Cooperative Extension Vegetable Crops Advisor, Riverside County
Eric B. Brennan, Organic Farming Research Scientist, USDA ARS, Salinas
Robert L. Bugg, Senior Analyst, Sustainable Agriculture Research and Education Program, University of California, Davis, Retired
Michael D. Cahn, University of California Cooperative Extension Irrigation and Water Resources Advisor, Monterey County
William E. Chaney, University of California Cooperative Extension Entomology Advisor Emeritus, Monterey County
Oleg Daugovish, University of California Cooperative Extension Vegetable Crops Advisor, Ventura County
Steven Fennimore, University of California Cooperative Extension Weed Specialist, U.S. Agricultural Research Station, Salinas
Howard Ferris, Professor, Department of Nematology, University of California, Davis
Mark Gaskell, University of California Cooperative Extension Specialty Crops Advisor, Santa Barbara County
Timothy K. Hartz, Cooperative Extension Vegetable Crops Specialist, Department of Plant Sciences, University of California, Davis
William R. Horwath, Professor, Soils and Biogeochemistry Program, University of California, Davis
Louise E. Jackson, Professor, Land, Air, and Water Resources, University of California, Davis
Karen M. Klonsky, Cooperative Extension Agricultural and Resource Economics Specialist, University of California, Davis
Steven T. Koike, University of California Cooperative Extension Plant Pathology Advisor, Monterey County
Jeffrey P. Mitchell, Cooperative Extension Cropping Systems Specialist, Kearney Agricultural Center, Parlier
Milton E. McGiffen Jr., Cooperative Extension Plant Physiology Specialist, Botany and Plant Sciences, University of California, Riverside
Gene Miyao, University of California Cooperative Extension Vegetable Crops Advisor, Yolo County
Joe Nuñez, University of California Cooperative Extension Vegetable Crops Advisor, Kern County
William H. Settle, Senior Technical Officer, UN World Food and Agriculture Organization, Rome
Johan Six, Assistant Professor, Department of Plant Sciences, University of California, Davis
Richard Smith, University of California Cooperative Extension Farm Advisor, Monterey, Santa Cruz, and San Benito Counties
Krishna V. Subbarao, University of California Cooperative Extension Plant Pathology Specialist, U.S. Agricultural Research Station, Salinas
Laura Tourte, University of California Cooperative Extension Farm Management Advisor, Santa Cruz County
Mark Van Horn, Director, Student Farm, University of California Davis

PART 1

Cover Crops for Vegetables and Their Uses

1 *Robert L. Bugg and Richard Smith*

Introduction

California is the leading vegetable crop–producing state in the United States. As of 2009, the top twenty-four vegetable crops in California were grown on 887,946 acres and had a value of over $5.4 billion (NASS 2010). Principal production areas include the coastal valleys, the southern desert, and the Central Valley. The vegetable industry is a complex mix of commodities that are grown for processing and the fresh market. Highly developed infrastructure enables vegetables to be shipped to wholesale markets across the United States and exported to other countries. An important volume of produce is also sold directly to consumers at local and farmers' markets, retail outlets, farm stands, and through community-supported agriculture.

Vegetable production has several characteristics that create problems for the long-term maintenance of soil quality and that increase the risk of off-farm movement of fertilizers and pesticides, which has negative environmental consequences.

- Intensive production on relatively high-value land limits the range of rotational crops (e.g., excludes agronomic crops such as wheat).
- Short, intensive cropping cycles make it difficult to include effective rotation crops in the production scheme.
- Shallow rooting systems and production practices increase the risk of nitrates leaching from the production system.
- Low amounts of biomass returned to the soil make the maintenance of soil organic matter levels and tilth challenging.
- Intensive tillage further increases the challenges of maintaining soil organic matter levels and tilth.

As a result of the above considerations, many growers are searching for cultural practices that can address and correct soil quality and environmental issues in vegetable production. This is especially important because growers are increasingly required to protect and improve the quality of water leaving their farmland. Cover crops provide an important tool for accomplishing these goals. Cover crops are not grown for harvest but are incorporated into the soil or left on the soil surface. They can

- reduce soil erosion by water or wind
- increase water infiltration
- fix atmospheric nitrogen that later becomes available to harvestable crops
- retain soil nitrogen and prevent it from leaching into groundwater
- enhance soil organic matter and improve long-term soil fertility
- break the life cycles of plant pathogens and other pests
- provide habitat for beneficial insects

Cover crops are grown mostly during the winter, when they cover and protect the soil from the force of raindrops and from wind erosion. Their roots bind the soil, and the organic matter they produce helps form water-stable soil aggregates that reduce soil erosion. Cover crop roots create macropores that facilitate air and water movement deep into the soil profile. The improved water infiltration often leads to decreased runoff. Grasses and mustard family plants can be used as "catch crops" to capture and retain soil nitrate in the upper soil profile, and plants in the pea family, if properly paired with their symbiotic *Rhizobium* bacteria, can fix enough atmospheric nitrogen so that, after plow-down, they can provide a significant proportion of the nitrogen needs of succeeding vegetable crops.

Cover crops are typically grown during fallow periods of the year, such as the winter in the northern part of the state or the summer in the southern desert. In this way they can be incorporated into the crop production system with a minimal impact on vegetable production schedules. However, because of the need to keep the land open and ready for spring planting, cover crops are typically grown on only a small percentage of vegetable acreage, even in the winter. One of the challenges for the future of vegetable farming in California will be to find ways to include cover crops, or at least their key functions, in a larger percentage of vegetable production systems.

Cover crops can be viewed as having complementary and overlapping roles with insectary plants, vegetative filter strips, tailwater ponds, hedgerows, and windbreaks for sustaining vegetable farming in California. These practices do not typically provide direct economic payback, but they do help address long-term soil production as well as the economic and environmental viability of vegetable production regions. These considerations have spurred recognition of and renewed interest in the ancient practice of cover cropping.

Since the mid-1980s, growers and scientists alike have explored a range of cover cropping options for vegetable farming systems, including those that include reduced tillage. Aiming for a synthesis of practical and scientific considerations, this publication reports some of their findings in chapters addressing botany and species selection; effects of cover crops on soil ecology; water management and impacts on water quality; soil nitrogen fertility management; pest management, including separate chapters on weeds, soilborne diseases, plant and soil nematodes, and arthropods; cover crop management; regional practices, including variations from the coastal, desert, and Central Valley regions; and economics.

References

NASS (National Agricultural Statistics Service). 2010. Vegetable 2009 summary. Washington, DC: USDA National Agricultural Statistics Service Agricultural Statistics Board.

2 *Richard Smith, Robert L. Bugg, and Oleg Daugovish*

Botany and Species Selection

Overview of Species

Vegetable crop production typically involves more tillage and higher nutrient use than other types of agronomic crop production, which affects selecting a cover crop species. Selecting a cover crop must be based on the growing region, production system, time of year, and the intended impact of the cover crop, as well as the amount of time and the season dedicated to cover crop production. This chapter discusses the primary cover crop species used in vegetable production in relation to these considerations.

Cover crop species fall into four main categories: grasses, legumes, mustards, and miscellaneous types (see tables 2.1 and 2.2).

Grasses

Grasses (Poaceae) include barley, oat, wheat, cereal rye, triticale, and sudangrass. These are used in vegetable production systems because of their fast initial growth, high biomass production, and low-priced seed. Grasses have fibrous root systems that make them good choices for capturing nitrate and preventing soil erosion. In the tops of grasses, the ratio of carbon to nitrogen (C:N) increases as the grasses mature. The carbon-rich residues of cereals also increase in lignin content as the crop matures and can be difficult to incorporate into the soil and slow to break down: high-carbon biomass may temporarily reduce available soil nitrogen as microbes use available soil nitrogen to decompose the residue (see chapter 5).

Cool-season grasses tolerate cold weather and grow vigorously during the winter in the principal vegetable production regions of the state. There are many different varieties of cereal grasses within a species; varieties range from winter-dormant to non-winter-dormant types and from short- to tall-growing types. Adapted varieties provide maximum performance. Grass cover crops differ in maturity date: early-maturing varieties may pose the hazard of setting seed if wet conditions persist late into the spring, making it difficult to mechanically incorporate them into the soil prior to seed set. Winter-grown grass cover crops can produce 9,000 to 11,000 pounds of dry biomass per acre (Mitchell et al. 2006). Among the available cool-season varieties, triticale produces the most biomass (L. Jackson, pers. comm.), while sudangrass is highly productive in the warm season.

Grass cover crops can be good choices for fields infested with plant-parasitic nematodes (see chapter 8) and are not hosts of Verticillium wilt (see chapter 7). They attract grain aphids (which are not pests of vegetables) that can help build up populations of aphid predators and parasites (see chapter 9).

Legumes

Legumes, members of the bean family (Fabaceae), can associate with soilborne bacteria in the genus *Rhizobium* and closely related genera, collectively known as rhizobia. In association with rhizobia, these plants form nodules on the roots and convert atmospheric nitrogen (N_2) to forms of nitrogen useful to the plant through the process of nitrogen fixation. As a result, legumes provide a net increase of nitrogen to the soil for crop production. The rate of nitrogen fixation peaks at the onset of flowering.

The seed of many legumes must be inoculated with the appropriate rhizobia strains at planting to ensure optimal nitrogen fixation (Philips and Williams 1987). Legumes vary in the amount of total nitrogen they produce, but their biomass contains a high percentage of nitrogen (3.5 to 5.0%), which allows their residues to break down rapidly upon incorporation into the soil (see chapter 5). Legumes such as vetches, bell beans, field peas, and cowpeas are often used as cover crops for vegetable operations. The total biomass produced by legumes can vary a great deal, but cover crop evaluations conducted from 1995 to 2000 on the Central Coast indicated that on

Table 2.1. Cover crop production and cultural information

Scientific name	Common name	Common varieties	Growing season*	Seeds per pound	Typical seeding rate[†] (lb/ac)	Seeding depth (inches)	Rhizobium inoculant	Seedling vigor	Vigor of winter growth[‡]
GRASSES									
Avena sativa	oat	Cayuse, Swan, Canota	cool	14,000–18,000	90–100	0.75–2.0	—	good	excellent
Hordeum vulgare	barley	UC 603, UC 937	cool	12,000–14,000	90–100	0.75–2.0	—	excellent	excellent
Secale cereale	cereal rye	Merced	cool	18,000–20,000	80–90	0.75–2.0	—	excellent	excellent
Sorghum bicolor × *S. bicolor* ssp. *drummondii*	sorghum-sudangrass	Sordan 79	warm	16,500	35–60	0.5–1.5	—	fair	none
Sorghum bicolor ssp. *drummondii*	sudangrass	Piper	warm	44,000	25–30	0.5–1.5	—	fair	none
Triticum aestivum	wheat	Summit, Dirkwin	cool	10,000–11,000	100–120	0.75–2.0	—	good to excellent	excellent
× *Triticosecale rimpaui*	triticale	Juan, Trios 102, Trios 348	cool	11,000–12,000	100–120	0.75–2.0	—	good to excellent	excellent
LEGUMES									
Crotalaria juncea	sunn hemp	Tropic Sun	warm	15,000	20–40	0.5–1.0	cowpea	good	—
Pisum sativum	field pea	Magnus, Biomaster, Dundale	cool	2,000–3,000	70–100	1.5–3.0	pea, vetch	fair to good	good
Trifolium spp. and *Medicago* spp.	clovers and medics	various	cool	110,000–170,000	15–25	0.25–0.5	various	poor	fair
Vicia benghalensis	purple vetch	various	cool	10,500	40–60	0.5–1.5	pea, vetch	fair to good	good
Vicia faba	bell bean	various	cool	1,200–1,500	125–175	1.5–3.0	pea, vetch	good	good
Vicia sativa ssp. *sativa*	common vetch	Languedoc, Vedoc, Willamette, Cahaba White, Vantage	cool	8,000	55–75	0.5–1.5	pea, vetch	poor	poor
Vicia villosa ssp. *varia*	woollypod vetch	Lana	cool	11,500	40–60	0.5–1.5	pea, vetch	fair to good	good
Vicia villosa ssp. *villosa*	hairy vetch	various	cool	16,200	30–50	0.5–1.5	pea, vetch	fair	poor
Vigna unguiculata ssp. *unguiculata*	cowpea	Iron Clay, Chinese Red	warm	3,000–4,000	40–60	1.0–1.5	cowpea, lespedeza	excellent	—
MUSTARDS									
Brassica juncea	Indian mustard	Pacific Gold, Caliente 99	cool	145,000–345,000	7–10	0.25–0.5	—	excellent	excellent
Brassica napus	canola	Humus, Erica	cool	90,000–120,000	7–10	0.25–0.5	—	good	
Raphanus sativus	oil seed radish	Nematrap	cool	40,000–50,000	10–20	0.25–0.75	—	excellent	excellent
Sinapis alba	white mustard	Ida Gold	cool	90,000–100,000	10–15	0.25–0.75	—	excellent	excellent
MISCELLANEOUS TYPES									
Fagopyrum esculentum	buckwheat	various	warm	20,400	20–30	0.5–1.5	—	excellent	—
Phacelia tanacetifolia	phacelia	Phaci	cool	220,000	10–20	0.5–1.0	—	good	good

Notes: *Some cover crops can grow out of their season, but yields will be less.
[†]Rates vary due to seed size and reasons for planting cover crop.
[‡]Cereal varieties differ in winter dormancy characteristics.

average legume cover crops produced about half the amount of biomass as grass cover crops (Smith 2011). Legumes tend to be more cold sensitive than other cover crops and require careful selection of varieties and planting dates. For instance, as a general rule in northern California, legumes should be seeded and germinated by mid-October to ensure good establishment and fall growth.

Many legume cover crops provide excellent habitat for beneficial insects. Bell beans, common vetch, and cowpeas have extrafloral nectaries that attract beneficial insects.

Mustards

Mustards (Brassicaceae) include Indian mustard, white mustard, canola, radish, and even some

Table 2.2. Characteristics of cover crops

Scientific name	Common name	Nitrogen fixation	Nitrogen scavenging	C:N ratio at maturity	Seeding hazard	Weed suppression	Comments
GRASSES							
Avena sativa	oat	—	excellent	high	low	moderate	Some varieties are the latest maturing of the cereals.
Hordeum vulgare	barley	—	excellent	high	high	good	Fast growing; many varieties are early maturing.
Secale cereale	cereal rye	—	excellent	high	moderate	excellent	At appropriate seeding rates, the variety Merced suppresses weeds well. AG 104 is resistant to rust.
Sorghum bicolor × *S. bicolor* ssp. *drummondii*	sorghum-sudangrass	—	excellent	high	moderate	good	Residues are allelopathic to tomatoes, lettuce, cole crops, and possibly other vegetables. Need to allow 6–8 weeks following incorporation of residue for allelochemicals to leach and/or break down.
Sorghum bicolor ssp. *drummondii*	sudangrass	—	excellent	high	moderate	good	Is tolerant of mowing and can produce large amounts of biomass.
Triticum aestivum	wheat	—	excellent	high	high	good	Its use as a cover crop is typically dependent upon seed price.
× *Triticosecale rimpaui*	triticale	—	excellent	high	moderate	good	Typically produces the most biomass of the cool-season grass cover crops. Resistant to rust.
LEGUMES							
Crotalaria juncea	crotalaria	excellent	poor	low	low	moderate	Upright growth to 6 feet tall.
Pisum sativum	field pea	excellent	poor	low	moderate	poor	Can be highly productive, but susceptible to root rot on heavy, wet soils.
Vicia benghalensis	purple vetch	excellent	poor	low	moderate	moderate	Tolerant of wet soils; a high nitrogen producer.
Vicia faba	bell bean	excellent	poor	low	moderate	poor	A reliably productive legume. Because of its upright growth it mixes well with grasses.
Vicia sativa ssp. *sativa*	common vetch	good	poor	low	moderate	poor	Poor winter growth and not a good choice for early plow-down.
Vicia villosa ssp. *varia*	woollypod vetch	excellent	poor	low	moderate	moderate to good	Early maturing; a high nitrogen producer.
Vicia villosa ssp. *villosa*	hairy vetch	good	poor	low	moderate	poor to moderate	Poor winter growth but grows vigorously in the spring. Grows well on sandy soils.
Vigna unguiculata ssp. *unguiculata*	cowpea	excellent	poor	low	low	good	Grows well in the extreme heat of the desert.
MUSTARDS							
Brassica juncea	Indian mustard	—	excellent	moderate	high	excellent	Small seeded, flowers later than white mustard.
Brassica napus	canola	—	excellent	moderate	high	good	Initially low growing. Late maturing.
Raphanus sativus	oil seed radish	—	excellent	moderate	high	excellent	Forms large, bulbous roots that can be difficult to incorporate.
Sinapis alba	white mustard	—	excellent	moderate	high	excellent	Vigorous upright growth.
MISCELLANEOUS TYPES							
Fagopyrum esculentum	buckwheat	—	excellent	moderate	high	good	Quickest maturing of the cover crops (i.e., 35 to 40 days to flower).
Phacelia tanacetifolia	phacelia	—	excellent	moderate	moderate	good	Residue is easy to incorporate into the soil.

vegetables. They contain glucosinolates, a class of chemicals that break down enzymatically to various short-lived chemicals such as isothiocyanates that have been shown in laboratory studies to control soilborne pathogens and weeds. There is interest in using mustard cover crops to biofumigate the soil and provide control of soilborne diseases and weeds, but to date, field studies have shown little impact in this regard (Bensen et al. 2009). Mustards are small seeded, but their vigorous seedling growth can outcompete weeds. Mustards can produce 6,000 to 9,000 pounds of biomass per acre. They have taproots and can absorb residual nitrate from deep in the soil profile; they are intermediate between grasses and legumes in the percentage of nitrogen in their tissue at maturity, and as a result they can cycle nitrogen quickly back to the soil. Even older stems tend to shatter when mowed and are easily incorporated into the soil. Standing mustard is used in some parts of the state to reduce wind erosion in fields.

All mustard and radish cover crops may host diamondback moth and flea beetles that normally do not reduce cover crop biomass production but can be pests of neighboring or succeeding cruciferous cash crops. Some mustard cover crop species are hosts of foliar diseases such as bacterial leafspot *(Pseudomonas syringae* pv. *alisalensis)* and soilborne diseases such as clubroot *(Plasmodiophora brassicae)* and white mold *(Sclerotinia minor)*; their use should be carefully considered in rotation with cash crops in the mustard family and other vegetables.

Miscellaneous Types

Many other species of plants can be grown as cover crops in vegetable crop rotations. The key issues are their growth patterns, seed price, availability, and desirable impact. Two notable species grown as cover crops are buckwheat *(Fagopyrum esculentum)* and phacelia *(Phacelia tanacetifolia)*.

Cover Crop Mixes

Growing mixes of two or more cover crop species can produce multiple benefits. Mixes of cereals and legumes combine the nitrogen-scavenging of cereals with the nitrogen-fixing of legumes. These mixes can maximize carbon and nitrogen inputs to the soil, enhance weed suppression, provide beneficial insect habitat, and reduce disease susceptibility. For instance, in one field study, a lettuce crop grown after a purple vetch-rye mix had less disease caused by *Sclerotinia minor* than a lettuce crop grown after a pure stand of purple vetch (Dillard and Grogan 1985). Growing multiple species can also ensure that at least one species will produce well under difficult conditions. Careful selection of species and relative seeding rates is necessary to minimize competition between mixture components. For instance, in order to reduce cereal competition with legumes, the cereal seeding rates must be reduced; unpublished data indicate that mustards are too competitive to mix with legumes.

Figure 2.1. Oat *(Avena sativa)*. J. K. Clark

Grasses

Cool-Season

Oat

Avena sativa

Figure 2.1

Most oat varieties mature later than do other species of cereals, but individual varieties may differ. Cayuse oat, one of the latest-maturing cereal varieties, is grown to reduce the risk of cover crop seed production in late, wet springs before equipment can incorporate the crop. Cayuse seedlings hold their first leaves upright and do not thoroughly shade the soil. This can encourage weed establishment, development, and seed set, although this problem can be reduced by increasing seeding rates.

Barley

Hordeum vulgare

Figure 2.2

Barley is a relatively large-seeded cereal that has rapid seedling growth. It is early maturing and is occasionally grown as a quick fall cover crop that is incorporated prior to the onset of winter. Barley may become a problematic weed if spring conditions do not allow

Figure 2.2. Barley *(Hordeum vulgare)*. E. Brennan

incorporation prior to seed set. Short and tall varieties are available. Barley is the most salt-tolerant cereal, but it is not as tolerant of heavy, wet soils as are triticale and cereal rye.

Cereal Rye

Secale cereale

Figure 2.3

The Merced variety of cereal rye is widely planted in California due to its excellent winter productivity and wide adaptability to conditions such as heavy, wet soils. It is moderately late maturing and has a high seed count per pound. Cereal rye is the most cold tolerant of the cereals and is thus useful for late-fall and winter plantings. Seedlings of cereal rye shade the soil early and provide excellent competition with weeds. It is susceptible to rust in prolonged wet winters; although rust does not reduce biomass production, it does cause early senescence of the leaves. The cereal rye variety AG 104 is resistant to rust and is also moderately late maturing. Cereal rye attracts grain aphids, which do not cause problems on vegetables; the presence of aphids encourages populations of aphid parasites and predators to build up early in the production season.

Wheat

Triticum aestivum

Figure 2.4

Wheat is occasionally used as a cover crop, especially if inexpensive seed is available. Numerous varieties are available that produce similarly to other cereal species. Common wheat varieties are intermediate in maturity.

Figure 2.3. Cereal rye *(Secale cereale)*. R. Smith

Figure 2.4. Wheat *(Triticum aestivum)*. C. A. Ingels

Figure 2.5. Triticale *(× Triticosecale rimpaui)*. E. BRENNAN

Figure 2.6. Sudangrass *(Sorghum sudanense)*. R. SMITH

Triticale

× Triticosecale rimpaui

FIGURE 2.5

Triticale is a cross between wheat and cereal rye. It is the most productive grass cover crop, but its use depends on the price of seed. The upright varieties such as Juan are used as cover crops for vegetables. However, the deep-rooted, winter-dormant variety Trios 102 may be used on furrow bottoms of fallow winter beds to reduce sediment loss and trap nitrate because it is low-growing until spring, when upright growth begins.

Warm-Season

Sudangrass

Sorghum sudanense

FIGURE 2.6

Sorghum-Sudangrass

*Sorghum bicolor (*syn. *S. vulgare) ×*
S. sudanense

Sudangrass and sorghum-sudangrass hybrids are warm-season grasses grown in the summer months. They tolerate the summer heat in the desert and are also grown as summer cover crops on the coast. They produce large amounts of biomass and tolerate mowing. They are small seeded, and their seedlings do not compete well with weeds. The residue is allelopathic to subsequent vegetable crops for 6 to 8 weeks; the allelochemicals must be leached with irrigation or rainfall prior to planting vegetable crops.

Legumes

Cool-Season

Bell Bean

Vicia faba

FIGURE 2.7

Bell bean is a small-seeded selection of fava bean, which makes it more economical for seeding as a cover crop. It differs from other true vetches by its upright central stem. This upright growth allows bell bean to be successfully grown in planting mixes with cereals. Bell bean alone, however, is a poor weed suppressor. Bell bean is more resistant to *Sclerotinia minor* than are other vetches, which is important in lettuce-producing areas of the Central Coast (Koike et al. 1996). It is susceptible to black bean aphid, which seldom affects its performance as a cover crop, but the aphid can vector viral diseases and be a direct pest of key vegetables such as beans and celery. Bell bean reliably produces moderate to high amounts of biomass and nitrogen. It is susceptible to Botrytis leaf spot in wet years, but in most cases, this disease does not appear to reduce biomass production.

Figure 2.7. Bell bean *(Vicia faba)*. J. K. CLARK

Purple Vetch

Vicia benghalensis

FIGURE 2.8

Purple vetch is a trailing plant that uses tendrils to climb. It is more tolerant of wet soils than are other vetches. It blooms later than woollypod vetch. In mild California winters it grows through the winter and can produce moderate to high amounts of biomass and nitrogen. At higher elevations with colder winter temperatures, it is sensitive to frost but will survive temperatures to 20°F for short periods. Its stems trail along the ground, and stems and leaves that touch the soil are susceptible to infection by *Sclerotinia minor.*

Woollypod Vetch

Vicia villosa ssp. *varia*

FIGURE 2.9

Woollypod vetch (Lana vetch) is a trailing plant that uses tendrils to climb. Its vigorous early growth is prostrate and moderately competitive with weeds. It is the earliest-flowering vetch. Woolypod vetch is more sensitive to wet soil than is purple vetch, and root rot can occasionally reduce stands on heavy, saturated soils. In mild California winters it grows through the winter and can produce moderate to high amounts of biomass and nitrogen. Its prostrate growth habit can form a mat of vegetation; the stems and leaves that touch the soil are susceptible to infection by *Sclerotinia minor.*

Figure 2.8. Purple vetch *(Vicia benghalensis)*. R. SMITH

Figure 2.9. Woollypod vetch (*Vicia villosa* ssp. *varia*). J. K. CLARK

Figure 2.10. Common vetch *(Vicia sativa* ssp. *sativa).* J. M. DiTomaso

Figure 2.11. Hairy vetch *(Vicia villosa* ssp. *villosa).* J. M. DiTomaso

Figure 2.12. Magnus pea *(Pisum sativum).* R. Smith

Common Vetch

Vicia sativa ssp. *sativa*

FIGURE 2.10

Common vetch is a trailing plant that uses tendrils to climb. It tends to stop growing even in the mild California winters and is therefore less productive than woollypod and purple vetches as an early plow-down cover crop. It is a poor competitor with weeds and is susceptible to *Sclerotinia minor*.

Hairy Vetch

Vicia villosa ssp. *villosa*

FIGURE 2.11

Hairy vetch (sand vetch) is a trailing plant that uses tendrils to climb. It is winter hardy and is commonly grown in the eastern United States. In cold locations, it can be planted before the ground freezes and remains dormant over the winter, resuming growth when the soil thaws in the spring. In mild parts of California it grows slowly during the winter, which makes it less productive as an early-plow-down cover crop, but if allowed to mature later in the spring it can be reasonably productive. It tolerates sandy soils.

Field Pea

Pisum sativum

FIGURE 2.12

Field peas are trailing to semiupright plants with tendrils that can climb. Many field pea varieties are available, and they differ greatly in winter dormancy and earliness; two examples include Austrian and Magnus. Austrian goes dormant during the winter and produces most of its biomass later in the spring; it is therefore not a good choice for early plow-down. Magnus grows actively in the mild California winters. Peas are large seeded with a relatively low seed count per pound; the variety Dundale was selected for smaller seed. In general, peas are sensitive to waterlogged soils, which limits their productivity on heavy soils in wet winters. Many pea varieties compete poorly with weeds and are moderately susceptible to *Sclerotinia minor*.

Clovers

Trifolium spp.

FIGURE 2.13

Medics

Medicago spp.

FIGURE 2.14

Clovers and medics are prostrate to erect plants. Their numerous species and varieties vary widely in productivity. They generally have low seedling vigor and lower total biomass production than vetches, peas, and bell beans. They compete poorly with weeds, and some species of clovers and most species of medics have hard seed and can thus become weeds. They are seldom used on vegetable ground in California, but species such as crimson clover *(Trifolium incarnatum)* are occasionally planted in vegetable operations and can be moderately productive when fully mature. Clovers are small seeded and must be carefully managed from seedbed preparation to the germination phase.

Figure 2.13. Clover *(Trifolium* spp.*)*. R. Smith

Figure 2.15. Cowpea *(Vigna unguiculata* ssp. *unguiculata)* seedlings. M. Van Horn

Warm-Season

Cowpeas

Vigna unguiculata ssp. *unguiculata*

Figure 2.15

Cowpeas are suberect to erect plants. Of the many varieties available, indeterminate types such as Iron Clay and Chinese Red are generally used as cover crops. They are adapted to heat and can be reasonably productive in as little as 60 days. They produce rapid, dense shade and compete with weeds. In the desert, their biomass has been used as mulch for subsequent transplanted vegetables. Many cowpea varieties are susceptible to root-knot nematode. Iron Clay possesses the Rk gene for root-knot nematode resistance, which makes it resistant to nearly all strains of *Meloidogyne incognita* but only moderately resistant to *M. javanica*.

Sunn Hemp

Crotolaria juncea

Figure 2.16

Sunn hemp is a fast growing, erect cover crop that can grow to 6 feet. It is adapted to heat and suppresses root-knot nematode.

Figure 2.14. Medic *(Medicago* spp.*)*. J. K. Clark

Figure 2.16. Sunn hemp *(Crotolaria juncea)*. R. Smith

Figure 2.17. *Brassica* spp. R. SMITH

Figure 2.18. White mustard *(Sinapis alba)*. R. SMITH

Figure 2.19. Radish *(Raphanus sativus)*. R. SMITH

Mustard Family

Cool-Season

Mustards

FIGURES 2.17 AND 2.18

The many species of mustards include Indian mustard *(Brassica juncea)*, white mustard *(Sinapis alba)*, canola *(Brassica napus)*, turnip *(B. rapa)* and black mustard *(B. nigra)*. They have high seed count per pound and require low seeding rates. Their vigorous seedlings compete well with weed seedlings. In most vegetable production areas of California they can be grown year-round. Winter plantings can be highly productive, but white mustard flowers early and can pose a risk of setting seed early in the spring. From late spring to early summer, mustards shift from vegetative to reproductive mode (bolt); plantings at this time often accumulate less biomass than do plantings from late summer to fall. All mustard cover crops may host diamondback moth and flea beetles that normally do not reduce cover crop biomass production but can build up and shift to neighboring or succeeding cruciferous cash crops. In addition, mustards are susceptible to foliar diseases such as bacterial leafspot *(Pseudomonas syringae* pv. *alisalensis)* and soilborne diseases such as club root *(Plasmodiophora brassicae)* and white mold *(Sclerotinia minor)* that can affect cash crops in the mustard family as well as other vegetable crops, and growing them as cover crops must be considered with care.

Radish

Raphanus sativus

FIGURE 2.19

The drawback to growing radishes as a cover crop is that the bulbous tap root may be difficult to incorporate and may regrow. This is less of a problem in freezing conditions, where the tap root rapidly disintegrates after thawing. Radishes are also susceptible to diamondback moth and flea beetles, as mentioned above for mustards. The varieties Nemagard and Maxi can suppress the cyst nematode.

Figure 2.20. Tansy phacelia *(Phacelia tanacetifolia)*. R. SMITH

Miscellaneous

Cool-Season

Tansy Phacelia
Phacelia tanacetifolia
FIGURE 2.20

Tansy phacelia is a native California annual wildflower introduced to Europe in 1910 as a nectar-producing plant for bees. Its feathery foliage and thin stems are easy to incorporate into the soil. It can absorb large quantities of residual nitrogen from the soil and cycle it quickly to subsequent crops. The available varieties were bred in Europe as a cover crop that is not a host to cyst nematode, thereby reducing populations of this pest prior to planting susceptible cash crops. It is susceptible to *Sclerotinia minor.*

Warm-Season

Buckwheat
Fagopyrum esculentum
FIGURE 2.21

Buckwheat is a fast-growing summer annual that is grown when a quick summer cover crop is needed. It flowers in as little as 35 days. The white flowers attract insect parasites and predators. Because it can set seed quickly, it must be incorporated into the soil promptly to avoid becoming a weed. Its biomass is easily incorporated into the soil and breaks down rapidly, quickly cycling nutrients to subsequent vegetable crops.

Figure 2.21. Buckwheat *(Fagopyrum esculentum)*. E. BRENNAN

References

Bensen, T. A., R. F. Smith, K. V. Subbarao, S. T. Koike, S. A. Fennimore, and S. Shem-Tov. 2009. Mustard and other cover crop effects vary on lettuce drop caused by *Sclerotinia minor* and weeds. Plant Disease 93(10): 1019–1027.

Dillard, H. R., and R. G. Grogan. 1985. Influence of green manure crops and lettuce on sclerotial populations of *Sclerotinia minor*. Plant Disease 69(7): 579–582.

Friedman, D. 1996. Evaluation of five cover crop species or mixes for nitrogen production and weed suppression in Sacramento Valley farming systems. Davis: University of California Cover Crop Research and Education Summaries 1994–1996.

Koike, S. T., R. F. Smith, L. E. Jackson, L. J. Wyland, W. E. Chaney, and J. I. Inman. 1997. Cover crops can increase lettuce drop. California Agriculture 51(1): 15–18.

Mitchell, J., M. Van Horn, D. Munier, and L. Jackson. 2006. Small grain cover crops. Small grain production manual, part 11. Oakland: University of California Agriculture and Natural Resources Publication 8174. ANR Communication Services Web site, http://anrcatalog.ucdavis.edu/CornGrains/8174.aspx.

Philips, D. A., and W. A. Williams. 1987. Range-legume inoculation and nitrogen fixation by root-nodule bacteria. Oakland: University of California Agriculture and Natural Resources Publication 1842.

Smith, R. F. In press. Cover crops for nutrient management in organic strawberry production on the Central Coast. In S. Koike, C. Bull, M. Bolda, and O. Daugovish, eds., Organic strawberry production manual. Oakland: University of California Agriculture and Natural Resources Publication 3531.

Summers, C. G., J. P. Mitchell, T. S. Prather, and J. J. Stapelton. 2009. Sudex cover crops can kill and stunt subsequent tomato, lettuce and broccoli transplants through allelopathy. California Agriculture 63(1): 35–40.

University of California Sustainable Agriculture Research and Education Program (UC SAREP) Cover Crop Resource Web site, http://www.sarep.ucdavis.edu/cgi-bin/ccrop.exe.

PART 2

Effects of Cover Cropping on Soil and Water Management

3 *Robert L. Bugg, William Horwath, and Johan Six*

Agricultural Soil Ecology

Agricultural soil ecology involves studying the interactions among soil components such as the physical matrix and its minerals; plant residues, leachates, exudates, and roots; soil microorganisms (e.g., bacteria, fungi, protozoa); meso- and macrofauna (e.g., nematodes, mites, insects, and earthworms); and grower practices, including techniques of tillage and irrigation. Knowledge of soil ecology can improve our ability to manage for crop production and environmental protection and can also enhance our insights into trade-offs between these goals. This chapter includes a brief primer on soil structure, organic matter, and biota, followed by a discussion of how cover cropping and associated organisms and processes can influence soil management in vegetable farming systems.

The organic substrates and living organisms of a soil can be viewed as parts of a complex food web. Organic matter is a reservoir for nutrients that sustain the growth of succeeding vegetable crops. Mineral particles (clay, silt, sand, and larger particles) make up about 50% of the total volume of the soil. The pores between these particles contain varying amounts of air and water and constitute most of the remaining 50% of the volume. In many nonagricultural settings, the soil food web is nutrient limited (oligotrophic), becoming more so with increasing depth. Vegetable farming systems often have soils that are, due to long-term mineral additions, rich in many nutrients and are limited only by relatively low availability of nitrogen and labile carbon. Additions of fresh organic matter through the growth and plow-down of cover crops lead to pulses of nitrogen, carbon, and various nutrients, activity by soil organisms, and changes in soil structure.

The Physical Matrix

Soil texture, the classification of a soil as clay, silt, sand, or loam, is determined by the proportions of primary soil particles (clay, silt, and sand) and is not alterable by cover cropping. By contrast, soil structure is the combination or arrangement of primary soil particles into secondary particles called aggregates, or peds. Aggregates can be categorized on the basis of size, shape, and degree of distinctness. Soil structure in the agricultural environment is affected by soil texture, mineralogy (mainly the types of clays and oxides), and organic and inorganic chemistry. Living organisms, mechanical field operations such as tillage, cover cropping, and irrigation also affect soil structure. In turn, soil structure affects root growth, the life cycles of soil microbiota and macrofauna, and the transport of soil gases, water, and chemical components, including plant nutrients and pesticides.

A soil aggregate is a composite of primary soil particles and organic matter. The soil organic matter occluded within aggregates is considered protected or physically isolated from microbial activity. This is an important type of carbon sequestration or storage that may reduce emissions of greenhouse gases, especially carbon dioxide. The protection is temporary, however, and soil aggregate dynamics are key to soil organic matter decomposition and stabilization, and therefore to soil fertility. Aggregate dynamics are driven by biomass additions, together with decomposer organisms, cycles of soil wetting and drying, and disturbances such as those caused by tillage (Six et al. 2004), and are major themes of soil science research. Indicators of degraded soil structure include aggregate instability, surface seals and crusts, high soil strength, impermeable subsurface horizons, and a reduction in the size, number, and continuity of macropores between aggregates. Some of these conditions can be managed, in part through the use of cover crops.

Soil aggregates are crumbs that consist of oriented clay and quartz particles. Water-stable soil aggregates do not collapse when wetted. Soils with abundant water-stable aggregates have reduced soil crusting and erosion and increased pore size, aeration, drainage, and nutrient retention. Macroaggregates and microaggregates are both important in soil dynamics. One macroaggregate may contain multiple microaggregates (fig. 3.1). Different forces operate at the different spatial scales. Fungal hyphae and roots are the main binding forces holding a macroaggregate together. Within a macroaggregate, microaggregates can form through the interaction of other forces that operate at a smaller spatial scale. Polyvalent cations are positively charged ions with the capacity for multiple charges. Calcium, for example, can behave as a polyvalent cation. Microaggregates are formed when polyvalent cations bind to the negatively charged clay surfaces. Polysaccharides derived from microbial activity form bridges between these polyvalent cations to cement the clay and silt particles together.

Microaggregates protect soil organic matter in the long term, whereas macroaggregate turnover is a crucial process influencing its stabilization (see fig. 3.1). Aggregate turnover takes place when an aggregate stabilizes around particulate organic matter encrusted with microbial products and earthworm mucus, then becomes unstable due to a cessation of microbial activity and eventually disrupts. Disturbances such as tillage enhance macroaggregate turnover, which diminishes the formation of new microaggregates within macroaggregates and the protection of soil organic matter in these microaggregates. Other things being equal, growers should reduce soil disturbance (e.g., tillage) as much as possible to reduce the breakup and consequently the turnover of macroaggregates. The reduced disturbance leads to a stabilization of soil organic matter in microaggregates that are themselves stabilized within slowly cycling macroaggregates.

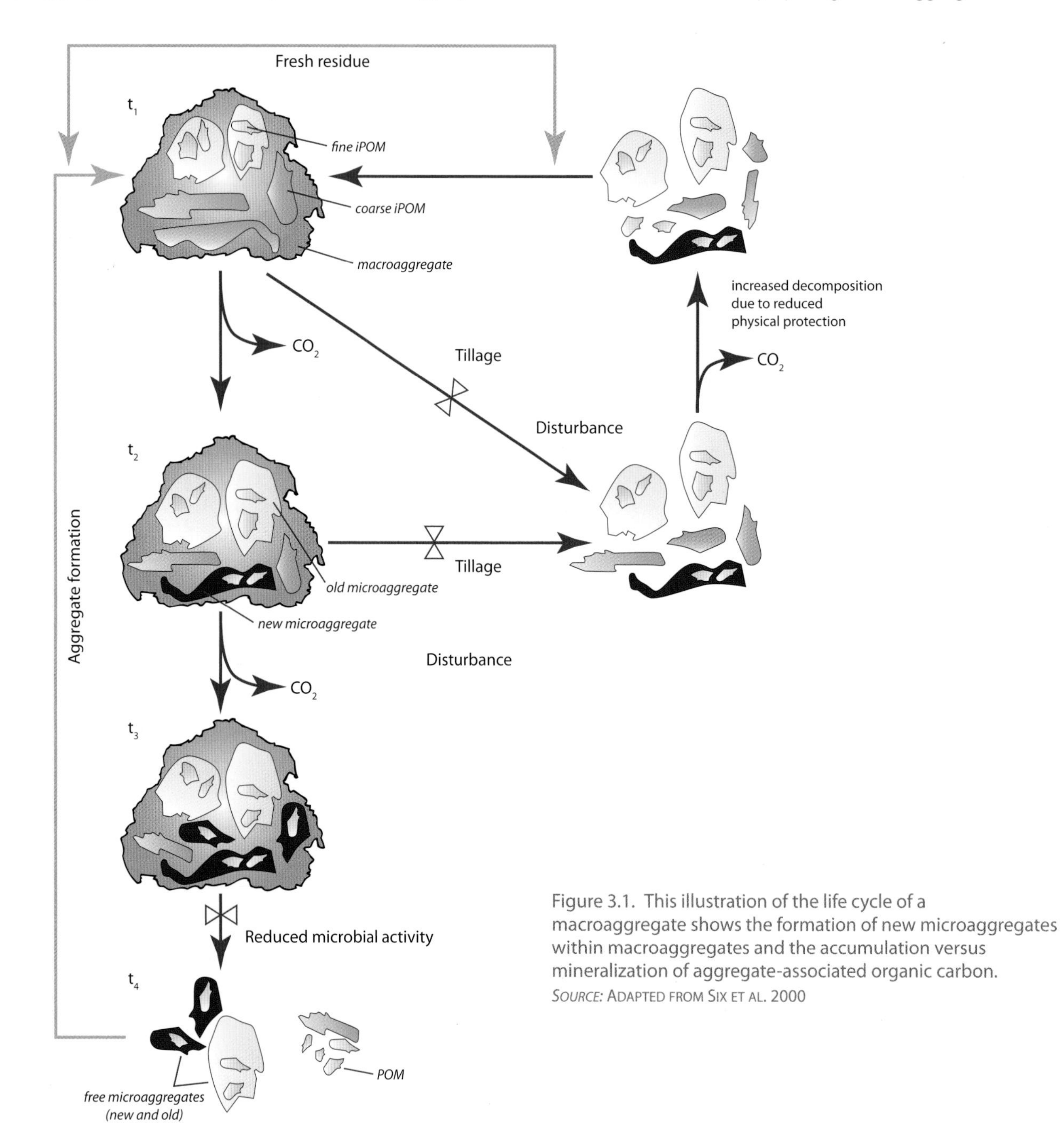

Figure 3.1. This illustration of the life cycle of a macroaggregate shows the formation of new microaggregates within macroaggregates and the accumulation versus mineralization of aggregate-associated organic carbon.
Source: Adapted from Six et al. 2000

The soil atmosphere is contained within and among the aggregate assemblages. A soil pore is a space or void between soil particles. Micropores (small pores) contribute to the water-holding capacity of a soil because their small size enables capillary forces to bind water tightly. By contrast, macropores (large pores) aid water transport in soils because their relatively large ratio of volume to surface area minimizes drag and capillary forces that impede gravity-driven fluid flow.

To illustrate the importance of macropores to water infiltration, a study of saturated flow in a fine sandy loam reported that a single macropore with a diameter of 3 millimeters (about 7 mm^2 area) in a soil area 30 centimeters in diameter contributed more to the steady infiltration rate than the remaining 71,000 square millimeters of the soil matrix (Smettem and Collis-George 1985). Macropores are formed by several processes, including root growth and burrowing by soil animals such as earthworms. These processes determine the characteristics of pores and their ability to function as networks that allow the movement of living organisms and the transport of air and soil gasses, water and solutes, and mineral and organic soil particles. Total soil porosity may be less important than size and continuity of pore space, which promote water infiltration and gas exchange. This was highlighted in a study in which conventionally tilled plots had greater total pore space but less pore continuity and lower rates of water infiltration than plots that were not tilled (McGarry et al. 2000).

Soil Organic Matter

A small fraction of the soil mass (typically 1 to 5%) is composed of soil organic matter (SOM) and includes nonliving material derived from plants, animals, and microbes, as well as living plant roots, soil microorganisms, and larger fauna. SOM performs important functions in agriculture. For example, it

- improves soil tilth by increasing the size and stability of soil aggregates, pore size, aeration, drainage, water-holding capacity, and resiliency (resistance to compaction)
- serves as a reservoir and source of soil nutrients
- provides sites for cation exchange
- chelates soil nutrients
- buffers soil pH

The formation and decomposition of SOM are fundamental processes that store and release energy and nutrients. SOM is composed mainly of carbon (55%) and up to 5 to 6% nitrogen. The phosphorus and sulfur contents of SOM are each about 1%. The formation of SOM is important in storing these nutrients.

The complex, variable structure and composition of SOM depend on complex chemical interactions with clay minerals. SOM is produced mainly by microbial and faunal decomposition of plant residues, root exudates, and waste from the trophic food web interactions. SOM is an important source of nitrogen, but the conversion of organic forms of nitrogen to forms useable by plants (the nitrogen mineralization process) is not well understood.

In California, agricultural soils have high decomposition rates that limit SOM accumulation, and these soils average between 1 to 3% SOM. Increasing crop residue inputs can maintain and even increase SOM in California in irrigated agricultural soils (De Clerck et al. 2003). However, most vegetable crops return low amounts of residues to the soil; therefore, cover crops play an important role in maintaining SOM (Mitchell et al. 1999).

Nonliving Components of Soil Organic Matter

The nonliving components of SOM can be categorized in several ways. One way is based on chemical composition, with the following three general categories:

1. Simple substances, including sugars, fats, amino acids, and other small molecules.
2. Identifiable high-molecular-weight organic materials such as polysaccharides and proteins.
3. Complex high-molecular-weight compounds or molecular colloids that cannot be fully characterized compositionally or structurally, including some polysaccharides and all humic substances.

Materials of categories 1 and 2 are direct inputs from crop and soil organism growth and turnover. In category 2, polysaccharides such as hemicellulose and cellulose are the largest plant carbon inputs to soil. They are chains of multiple sugar molecules composed of monomer units of simple sugars that are produced by plants or microbes. Cellulose and hemicellulose are structural polysaccharides that are components of plant and microorganism cell walls (fig. 3.2). The polysaccharides are enmeshed with lignin and proteins to create a rigid structure that supports the plant and wards off entry by pathogens. Their chemical compositions and structures are standardized and can be specified. Another subset of polyscaccharides comprises the sticky substances that are exuded by the roots of many plants including cover crops and by soil-dwelling bacteria. Extracellular enzymes degrade

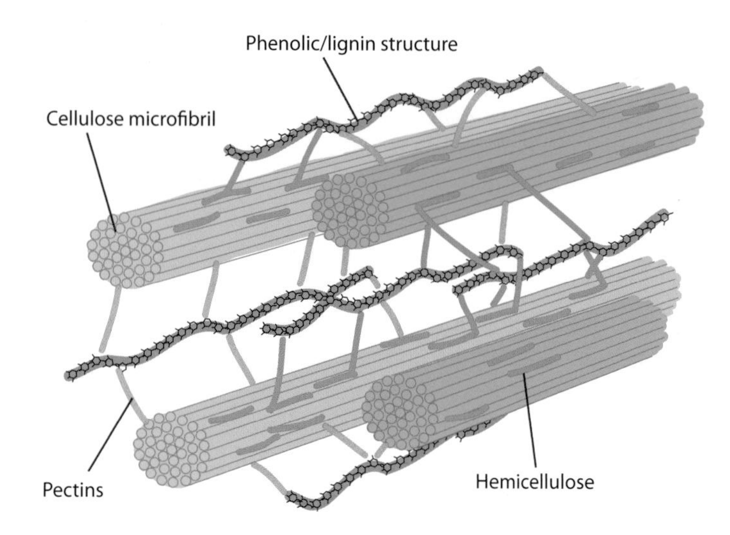

Figure 3.2. This three-dimensional model of the secondary cell wall of plants allows visualization of interactions among petic substances, hemicellulose, cellulose, lignin, and proteins such as extensin. *Source:* After Horwath 2007.

polysaccharides into simple sugars, a major food source supporting the soil food web (Martin and Haider 1971). Polysaccharides assist in binding small soil particles together (aggregate formation), typically after at least one cycle of wetting and drying. Water-stable aggregates, which do not collapse when wetted, reduce soil crusting and erosion while increasing pore size, aeration, drainage, and nutrient retention.

Category 3 compounds include polysaccharides and humic substances produced or modified extracellularly by soil microorganisms. These compounds differ from category 2 compounds in that their compositions and structures vary greatly among individual molecules. This occurs in part because these molecules are highly reactive with one another and with minerals, leading to the formation of chemical complexes.

In contrast with polysaccharides, humic compounds are formed primarily of aliphatic substances linked to phenolic and benzoic acid subunits that are derived from lignin, a structural compound from plants and fungi, and from other phenolic compounds such as melanins, which are produced by fungi and other soil microorganisms (Martin and Haider 1971; Stevenson 1994; Adani and Ricca 2004). Three subcategories of humic substances have been proposed:

- fulvic acids: soluble in both acid and alkaline solutions
- humic acids: soluble only in alkaline solutions
- humin: insoluble, often tightly bound to clay

The classical extraction of humic substances (fig. 3.3) is done using a strong alkali such as sodium hydroxide (NaOH) to dissolve the humic substances. The extractant is then separated by centrifugation, and a strong acid such as hydrochloric acid is added to separate fulvic from humic acids. Fulvic acids remain in solution while heavier humic acids precipitate under low pH. The nonextractable fraction, humin, is strongly associated with the mineral soil. The functions of these operationally defined fractions have been difficult to elucidate, but their quantities and turnover have been used to describe carbon and nutrient turnover in many agricultural systems.

Humic substances are important in maintaining soil fertility, especially on sandy soils with low nutrient availability. As noted by Stevenson (1994), humic substances assist with

- cation exchange
- chelation
- pH buffering
- physical resiliency (resistance to compaction)
- water-holding capacity

Cation exchange refers to the reversible binding of positively charged ions, including nutrients, which reduces their leaching through the soil profile. Clay minerals and humic molecules provide cation exchange sites; clay is less efficient per unit mass than are humic acids. However, although humic materials have a high cation exchange capacity, they represent a very small part of the soil mass. Humic substances are not an

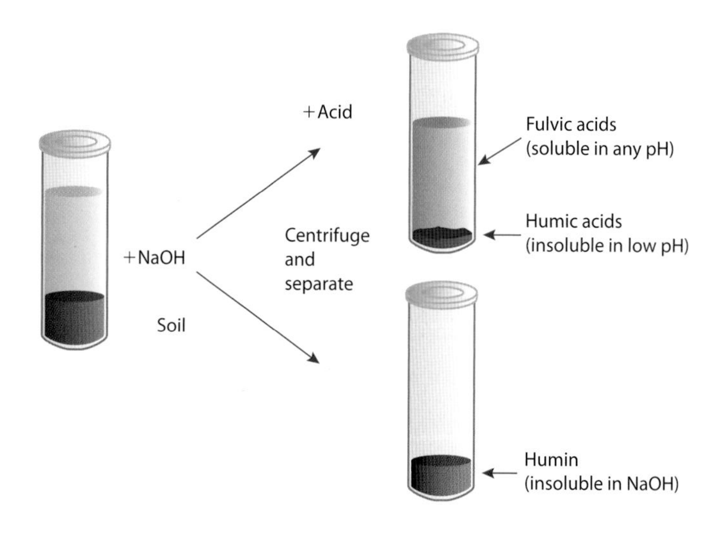

Figure 3.3. The classical extraction of humic substances.

important determinant of cation exchange capacity in soils with abundant clay, but they are very important in sandy soils.

SOM can also be classified based on molecular weight. When humic substances interact with and link to soil minerals such as clays, they become heavy. The heavy fraction of SOM constitutes 70 to 80% of the nonparticulate organic matter in most soils; this material, the heavy fraction organic matter, is very persistent through time. The large, amorphous assemblies of heavy-fraction organic matter are spongelike in architecture, containing spaces (voids) that are hydrophobic (water repellant). Some pesticides are adsorbed in these hydrophobic domains, leading to reduced pesticide activity. These heavy, stable humic substances can range from a few years to thousands of years in age.

The remaining lighter portion of SOM, the light fraction or particulate organic matter (POM), is composed of recently deposited material as well as older plant inputs protected in aggregates. The term light fraction is used because the material can be isolated from soil based on its buoyancy in a heavy liquid. Typically, the light fraction constitutes up to 30% of the SOM by mass, and it is a ready source of energy and nutrients for soil food web interactions. The light fraction responds quickly to management: changes in nutrient availability can often be realized within 1 to 3 years. The use of cover crops can greatly influence the size and function of the light fraction of SOM.

Living Components of the Soil

Roots, bacteria, fungi, and fauna compose the living soil biomass. Cover crop incorporation into the soil initiates an array of trophic interactions, releasing nutrients, mainly nitrogen, and making them available for uptake by subsequent vegetable crops. Soil organisms represent a small part of the total soil volume and mass but play a large role in the ecology and fertility of soils (table 3.1). The living components of SOM include many species that vary widely in function. Up to one million species of organisms, including bacteria, fungi, and animals (fauna), inhabit 1 gram of soil (Horwath 2002). This great diversity of soil life influences ecosystem processes of energy flow (the carbon cycle) and nutrient dynamics that in turn control plant growth.

The composition, structure, and function of the soil food web are influenced by the quantity and quality of plant residues entering the soil (Moore and Hunt 1988; Moore et al. 2004). Additions of crop residue, tillage disturbance, and application of pesticides also influence soil ecology, directly and indirectly harming some organisms while enhancing others.

Table 3.1. Soil biomass components

Soil biomass component	Tons per hectare
earthworms	< 2.5
fungi	2–5
bacteria	1–2
actinomycetes	1–2
protozoa	< 0.5
other fauna (collembola, mites, arthropods, etc.)	< 0.5
nematodes	< 0.2

Source: Adapted from Horwath 2002.
Note: 1 metric ton/hectare = 892 pounds per acre.

The diverse players in the soil food web perform a variety of functions. Large animals (macrofauna) such as earthworms function as detritivores, breaking down litter into smaller particles. Macrofauna can play important roles in carbon cycling and other soil processes, especially in vegetable farming systems with reduced tillage.

Smaller decomposers, including mutualistic bacteria in the intestines of detritivores, chemically degrade litter, releasing nutrients. Predators, parasites, and pathogens attack and digest other organisms, releasing more nutrients and controlling soil community composition and population densities. These predators, parasites, and pathogens provide feedback loops that regulate nutrient release. Less diverse systems may favor one or another of these classes, leading to less efficient nutrient cycling. The diversity of organisms in the three classes provides mutual back-up of ecosystem processes through functional redundancy.

The organisms of the soil food web have varied energy and nutrient demands and different carbon-to-nitrogen (C:N) ratios, leading to inefficiencies of conversion that allow release of plant-available nitrogen and other nutrients. The actions of these organisms, together with their byproducts, excreta, and dead bodies, enable formation and turnover of SOM, mediate the availability of nutrients, and change soil structure during and after crop growth. The combining of individual soil particles into larger, more complex physical units or aggregates creates an immense array of microhabitats, or niches, that delineate boundaries among the members of the trophic food web. They also constitute protected areas to which larger organisms and predators do not have access, buffering population fluctuations and enabling rapid recolonization.

The abundance of soil organisms and the fractions of total soil biomass they represent vary with crop, crop developmental stage, season, depth, degree of soil disturbance, and rotational history. At peak crop development, living roots are the largest biomass component in the top 6 to 12 inches of the soil. With greater depth, bacteria and fungi become relatively more important, even though their absolute biomass is lower than in the surface stratum. Among the microbial components, bacteria dominate soils that are frequently tilled, with fungi increasingly important as tillage is reduced (Horwath 2002; see table 3.1). Since bacteria and fungi represent most of the soil microbial biomass, they are considered responsible for the majority of energy flow and nutrient cycling.

Protozoa, nematodes, and amoebae influence nutrient cycles by grazing on bacteria and fungi, influencing the size of their populations. Bacteria and fungi have lower C:N ratios than do these grazers; by their consumption, excess nitrogen is released and mineralized to ammonium. At higher trophic levels, larger invertebrates such as mites, springtails (Collembola), and earthworms graze on protozoa, nematodes, and amoebae, releasing more nitrogen, which becomes available for plant uptake.

Springtails (see chapter 9) are small insects that are important in breaking down crop residues and controlling some plant root pathogens (Lartey et al. 1994). However, their role in crop plant vigor and yield appears mixed, because they may feed on crop plant tissues as well (Scheu et al. 1999). In California, one species, *Entomobrya unostrigata*, is a minor pest of vegetable crops.

Most studies indicate that earthworms benefit plant growth (Scheu 2003). Earthworm species vary greatly in their burrowing and feeding behaviors. Geophagous (earth-feeding) earthworms ingest soil that contains bacteria, fungi, and larger fauna such as protozoa, nematodes, dead plant roots, and litter. Earthworms are scarce in heavily tilled soils that receive little organic matter (Springett et al. 1992). The most common earthworms in California vegetable fields include *Allolobophora chlorotica, Aporrectodea caliginosa, Ap. longa, Ap. rosea,* and *Ap. trapezoides,* all of which are native to Eurasia (S. Fonte, pers. comm.). *Microscolex dubia*, which is native to South America, is also common. All these species are members of a complex called peregrine (wandering) earthworms that have been widely transported by humans (Lee 1987). The species listed above are mainly subsurface-feeding (endogeic) species that feed on decomposing plant material and soil containing abundant organic matter. *Aporrectodea caliginosa* can feed at the surface as well as below it; such species can adapt to different tillage systems and placements of crop residues in different depths in the soil profile (Cook and Linden 1996). Surface-living (epigeic) earthworms are rare in vegetable culture (Didden 2001), and, in California, deep-burrowing species that feed on the surface (anecic types, such as the nightcrawler, *Lumbricus terrestris*) are common only in the far northwestern part of the state.

Ecosystem engineering is the network of soil pores and changes in soil properties produced by earthworms and other fauna. Some endogeic species burrow extensively at depth, yet their burrows may seldom reach the soil surface and may also lack continuity because the worms refill the burrows with castings (excreta) (Hirth et al. 1996). These partially refilled, discontinuous burrows probably have less impact on water infiltration than do continuous channels.

Cover Crop Influence on Soil

Improved soil tilth and water infiltration are often attained after 1 or 2 years of cover cropping, whereas improved nutrient cycling is usually seen after 3 or more years. Cover crops provide biomass that serves as food for soil organisms, which in turn provide food for other organisms, thus supporting a food web with many trophic (feeding) levels. This biomass, associated organisms, and their breakdown products drive changes in soil structure, ecology, and nutrient dynamics. Cover crops add biomass to the soil through at least four processes: shedding of old foliage, root exudation, root biomass production, and incorporation (plow-down) or surface placement of cover crop tops (stems and leaves). The shedding of old foliage by cover crops has not been well studied, but it is probably an important source of biomass in vetches, field pea, and other covers.

Soil Structure

Cover crops affect soil structure through cover crop above- and below-ground growth patterns, their associated organisms (including decomposers), their exudates and associated decomposition products, and the agronomic practices used in their management. A growing cover crop provides a leafy canopy that protects against rain (or irrigation) splash and consequent translocation of soil particles, lessening the tendency toward sealing, crusting, and erosion.

Cover cropping enhances soil porosity, resulting in greater infiltration of rainfall (Connolly et al. 1998). Plant roots also create macropores, especially the roots of grasses, which are rich in carbon and contain abundant lignin, so that they decompose slowly (Williams 1966; Merwin et al. 1996; Miller et al. 1999; Ghodrati et al. 1999; Devitt and Smith 2002). These pores tend to be seasonal and depend on plant growth.

Macropores are also formed through the burrowing actions of earthworms and other decomposers that feed on root exudates, dead roots, and dead leaves, whether tilled under or left on the surface.

The influence of cover crops depends on the interaction of soil texture (clay, silt, and sand proportions) and mineralogy (mainly the types of clays and oxides) with SOM. Enhanced aggregation resulting from additions of fresh organic matter is seen more in fine-textured soils because the greater surface area presented by small primary particles renders them more amenable to bonding into aggregates.

Cover cropping also supports the formation and persistence of water-stable soil aggregates by polysaccharide gums, which are exuded by roots, or by polysaccharides produced as breakdown products of bacteria, which bind primary soil particles together. Fungal hyphae are also important in binding primary particles into macroaggregates. Geophagous (earth-feeding) earthworms play a paradoxical role by alternately destroying and remolding soil macroaggregates. These are produced as excreta (casts) that are formed within the earthworms' intestines by the application of pressure and the exudation of polysaccharide mucus (Bossuyt et al. 2004). In this process of destroying and remolding aggregates, earthworms digest part of the SOM but render other parts less accessible to organisms, thus enabling the storage of carbon in microaggregates that are formed within the macroaggregates.

Cover crops promote aggregate stability by adding fresh organic matter. In the Sustainable Agriculture Farming Systems project at UC Davis, cover crops yielded disproportionately higher soil carbon sequestration rates than expected based solely on the amount of carbon in the cover crop (Kong et al. 2005). This higher stabilization may be related to the residue quality (low C:N ratio) and root morphology (many fine roots) of cover crops (Kong et al. 2005). Furthermore, the mean residence time of soil aggregates (a measure of how long they persist) increased under cover crops (Kong et al. 2007). Earthworms in combination with cover crops enhance the incorporation of carbon and nitrogen into soil aggregates (Fonte et al. 2006). All these processes may lead to a preferential stabilization of cover crop carbon in the soil.

The use of legume cover crops, such as bell bean, field pea, hairy vetch, etc., in California cropping systems has consistently shown benefits in sequestering soil carbon and enhancing soil aggregate stability, water infiltration, and control of some pathogens (Horwath et al. 2002; Veenstra et al. 2008). Over a 10-year period, up to 5 metric tons of carbon per hectare can be sequestered using winter cover crops (Horwath et al. 2002). The soil property most affected by increasing SOM is water infiltration, with rates being enhanced almost twofold under cover crops. The storage of soil nitrogen as SOM and its increased availability to succeeding cash crops also increases dramatically with the use of winter cover crops (Poudel et al. 2001).

Ecology

The management of SOM requires continued inputs of crop residues or other organic amendments such as manure or compost. The end products of the initial decomposition of these inputs are required to maintain the level of SOM. This is especially important for organic systems that rely on SOM to provide a mineralizable pool of nutrients. The use of cover crops, both leguminous and cereal, is an excellent way to maintain SOM and nutrients in cropping systems. The carbon in SOM can play a role similar to that of fresh plant residue; it can also function as an energy source for the soil biota, and the decomposition of SOM releases nutrients for uptake by vegetable crops. Generally, 2 to 5% of SOM decomposes annually. Although the amount of nitrogen mineralized from incumbent SOM is in theory sufficient to meet the needs of some vegetable crops, the timing of release may not coincide with uptake requirements without management of the mineralization processes of the soil food web (see Ferris et al. 2004). The decomposition of crop residue and other amendments combined with the turnover of SOM provides nutrients for the soil food web and for plants. Overall, the long-term effect of cover cropping and organic amendments is to increase the nutrient-supplying capacity of the soil.

The value of winter cover crops in building SOM depends on the quality of the cover crop residue. Species selection is important because different cover crops have differing chemical compositions (e.g., cellulose or lignin contents) and C:N ratios (Quemada and Cabrera 1995), affecting their decomposition rate and potential to produce SOM. Increasing the C:N ratio and the level of more-resistant carbon-rich compounds such as lignin leads to slower decomposition (see tables 3.2 and 5.1). Consult chapter 5 for more details on nitrogen availability.

The maturity of cover crops affects their chemical composition: cover crops in the vegetative stage have a lower C:N ratio than do cover crops in the flowering

Table 3.2. Effect of C:N ratio (dry mass) on nitrogen availability

C:N ratio	Nitrogen availability
< 10	high
10–20	medium
20–30	low
> 30	negative

and seed development stages. The main determinant of initial decomposition rate is the cover crop C:N ratio. The lower the C:N ratio, the higher the decomposition rate. Therefore, residue of vegetative-stage crimson clover decomposes faster than residues from the flowering or early seed maturation stage (Ranells and Wagger 1992). To enhance cover crop nitrogen availability, the crop should be cut and incorporated into the soil at the vegetative stage.

The recovery of cover crop nitrogen by vegetable crops also depends on environmental factors (e.g., climate, soil temperature, and water conditions) and type of management (e.g., shredding, mixing). Fine chopping enhances the rate of decomposition and nutrient release. Uniform spatial distribution of cover crop residue accelerates aerobic decomposition and makes nitrogen available more uniformly to the vegetable crop. By contrast, clumps of cover crops, whether incorporated or left on the soil surface, may create anaerobic conditions and related phytotoxic compounds, encourage the development of pathogens and other pests, and produce erratic availability of nitrogen and other nutrients to crop plants, leading to uneven stand development and lower crop quality.

Soil and crop management can greatly impact SOM. For example, soil tillage and simplification of crop rotations are suspected to have led to as much as a 50% decrease in soil carbon compared with the native systems (Lal et al. 1998). Soil tillage reduces SOM by fracturing aggregates and exposing previously protected SOM to oxygenation. Tillage also differentially affects various soil organisms and their interactions, impacting the flow of energy and nutrients through the food web. Cover crops can somewhat mitigate these effects by increasing soil carbon, soil biodiversity, and food web function.

Nutrient Dynamics

The amount of nitrogen liberated by SOM depends in part on the nutrient demand of the vegetable crop, because plant roots exude carbon-rich compounds that stimulate soil microbial activity, leading to the breakdown of more nitrogen-rich SOM and mineralization of nitrogen. Plant root growth further enables this by physically disrupting soil aggregates and exposing the associated SOM to breakdown.

Legume-rich cover crops can add up to 133 pounds of nitrogen per acre in California cropping systems, depending on plant species, density, and growth (Poudel et al. 2001). Leguminous cover crops may contain 2.0 to 5.0% nitrogen (dry mass) at peak flower, based on dry matter analysis. Bell bean *(Vicia faba)* and annual clovers (*Trifolium* spp.) are on the low end of this continuum, and vetches (*Vicia* spp.) and field pea *(Pisum sativum* spp. *arvense)* are on the high end. Warm-season legumes such as cowpea *(Vigna unguiculata* ssp. *unguiculata)* range from about 2.0 to 3.0% (Harrison et al. 2004). By contrast, most cool-season cereal grains are about 1.0 to 1.5% nitrogen at peak flower. The C:N ratios of plant residues depend on nitrogen content, since the concentration of carbon varies little in plants. Winter annual legumes usually have a C:N ratio of less than 25:1, whereas grasses and many other forbs with lower nitrogen contents have ratios greater than 20:1, ranging as high as 80:1 if the crop is allowed to set seed. Residues of winter annual legumes decompose quickly, leading to nutrient mineralization for uptake by vegetable crops. If obtained at similar plant developmental stages, cereal residues are slower to break down than are legume residues.

Mustard family plants (Brassicaceae), including various *Brassica* species, are more efficient than grasses or legumes in taking up residual soil nitrate, partly because of their more rapid vertical root growth (Vos et al. 1998). Mustard family plants may have nitrogen contents around 3.5% by dry mass. The aboveground parts of white mustard *(Sinapis alba)* have a low C:N ratio, about 9 for leaves and 19 for stems, and a low lignin-to-nitrogen ratio, about 7 for leaves and 14 for stems (Chaves et al. 2004). Therefore, mustard crop residues break down rapidly in the soil, and, compared with grasses, liberate a surprisingly large amount of nitrogen, considering that mustard family plants also do not fix atmospheric nitrogen.

During the first year after cover crop establishment, only a small portion of the current-year cover crop nitrogen will become available to the succeeding vegetable crop. Factors affecting nitrogen availability during the early years of cover crop management include immobilization of nitrogen by growing soil organism communities. During subsequent years, nitrogen availability increases as a result of increasing SOM. Current-year nitrogen recovery rates in crops following legume cover crops range from 10% to over 50% of the cover crop nitrogen (Hadas et al. 2002).

The total amount of nitrogen mineralization from cover crop residues is inversely correlated with C:N ratio, regardless of how long cover crops have been in use. In addition, the increase in lignin content of cover crops in the flowering and seed development stages decreases the rate of nitrogen mineralization (Chaves et al. 2004). Table 3.3 presents the approximate amounts of nitrogen expected for a succeeding warm-season vegetable crop after incorporation at peak flower of a

Table 3.3. Approximate net annual soil nitrogen supplied by a legume/cereal mix cover crop to a warm-season vegetable crop that required 150 pounds of nitrogen per acre per year

	Total N supplied in year (lb/ac/year)			
Soil type	**0**	**1**	**2**	**3**
sand	60–75	60–80	65–85	75–95
loam	80–100	90–110	100–120	110–120
clay	70–90	75–95	85–105	90–110

Source: Adapted from Havlin et al. 2005.
Note: In year 0, no winter annual cover crop preceded the summer vegetable crop. In years 1, 2, and 3, a vetch-pea-bell bean-oat cover crop was plowed down containing 150 pounds of newly fixed nitrogen per acre with a C:N ratio of 25.

cool-season legume-and-cereal cover crop that has fixed 150 pounds of nitrogen per acre and has a composite C:N ratio of 25. Loam soils receive the greatest nitrogen benefit from cover cropping and sandy soils receive the least, with clay soils being intermediate (Havlin et al. 2005).

Cover crops also accumulate important amounts of phosphorus, potassium, and micronutrients in organic form, increasing their availability to succeeding crops. The accumulation of these nutrients in plant biomass keeps them in an active cycling state through decomposer activities and also makes them less prone to loss from runoff and leaching. Buckwheat *(Fagopyrum esculentum)* is grown as short-term (30- to 45-day) summer cover crop by many organic growers. It shows a high capacity for phosphorus uptake on alkaline soils, in part through exuded enzymes (Amann and Amberger 1989) and in part through acidification of the rhizosphere (Zhu et al. 2002). Buckwheat is often recommended for rotation on soils with limited phosphorus availability, because the assimilated phosphorus becomes available to vegetable crops after incorporation of the buckwheat residues.

Cover crop strategies have been developed to make nutrients readily available and maintain SOM. Legumes are often planted in mixtures with grasses because of the complementary functions of the plants, that is, nitrogen fixation versus nitrate scavenging. Mixing cover crops to enhance soil function and quality can often be related to C:N ratio (see table 3.2). When mixed cover crops are plowed into the soil, the legumes decompose rapidly, releasing nitrate that in turn assists in decomposing carbon-rich compounds such as simple sugars, starches, cellulose, hemicellulose, and lignin produced by the grasses. Thus, residues from a combination of legumes and grasses break down faster than those of grasses alone. The concept of mixing residues with varying C:N ratios can be applied to a variety of organic amendments and cropping systems. For more information on nutrient dynamics, see chapter 5.

Conclusion

The effects of cover cropping on soil ecology are mediated by soil type, cropping history, cover crop selection, temporal and spatial plant growth patterns, associated organisms, weather, and management regimes. These vary profoundly among sites and through time. No predictive model can take these variables into account. For example, nitrogen availability to a succeeding vegetable crop through time depends on many factors, including cover crop quality, time of incorporation, and soil temperature. Perhaps in the long run such models will become available, but they will need to include soil temperature and moisture to describe the decomposition process accurately and precisely.

The main foci of cover crop research have been on the direct effects of soil texture and fertility on the value of cover crops and the cover crop's impact on succeeding crop growth and yield. However, studies often address a specific component of cover crops, such as the role of decomposer macrofauna and microflora or nutrient value to succeeding crops. It is extremely important to know the interactions among cover crop growth, macrofauna, microflora, soil structure, and soil nitrogen availability in order to manage cover crops correctly for optimal crop yield. Furthermore, macrofauna can function as herbivores, fungivores, or as a food base for predators that attack crop pests. Soil macrofauna also serve as prey for birds and other wildlife, some of which can be tolerated in some vegetable farming systems. In light of this complexity, integrative research efforts are needed to elucidate the value of cover crops in vegetable crop production.

References

Adani, F., and G. Ricca. 2004. The contributions of alkali soluble (humic acid-like) and unhydrolyzed-alkali soluble (core-humic acid-like) fractions extracted from maize plant to the formation of soil humic acid. Chemosphere 56:13–22.

Amann, C., and A. Amberger. 1989. Phosphorus efficiency of buckwheat (*Fagopyrum esculentum*). Zeitschrift für Pflanzenernährung und Bodenkunde 152:181–189.

Bossuyt, H., J. Six, and P. F. Hendrix. 2004. Rapid incorporation of carbon from fresh residues into newly formed stable microaggregates within earthworm casts. European Journal of Soil Science 55:393–399.

———. 2005. Protection of soil carbon by microaggregates within earthworm casts. Soil Biology and Biochemistry 37:251–258.

———. 2006. Interactive effects of functionally different earthworm species on aggregation and incorporation and decomposition of newly added residue carbon. Geoderma 130:14–25.

Chaves, B., S. DeNeve, G. Hofman, P. Boeckx, and O. Van Cleemput. 2004. Nitrogen mineralization of vegetable root residues and green manures as related to their (bio)chemical composition. European Journal of Agronomy 21:161–170.

Connolly, R. D., D. M. Freebairn, and M. J. Bell. 1998. Change in soil infiltration associated with leys in south-eastern Queensland. Australian Journal of Soil Research 36:1057–1072.

Cook, S. M. F., and D. R. Linden. 1996. Effect of food type and placement on earthworm *(Aporrectodea tuberculata)* burrowing and soil turnover. Biology and Fertility of Soils 21:201–206.

De Clerck, F., M. J. Singer, and P. Lindert. 2003. A 60-year history of California soil quality using paired samples. Geoderma 114:215–230.

Devitt, D. A., and S. D. Smith. 2002. Root channel macropores enhance downward movement of water in a Mojave Desert ecosystem. Journal of Arid Environments 50:99–108.

Didden, W. A. M. 2001. Earthworm communities in grasslands and horticultural soils. Biology and Fertility of Soils 33:111–117.

Dominguez, J., P. J. Bohlen, and R. W. Parmelee. 2004. Earthworms increase nitrogen leaching to greater soil depths in row crop agroecosystems. Ecosystems 7:672–685.

Emerson, W. W. 1959. Stability of soil crumbs. Nature 183:538.

Ferris, H., R. C. Venette, and S. S. Lau. 1997. Population energetics of bacterial-feeding nematodes: carbon and nitrogen budgets. Soil Biology and Biochemistry 29:1183–1194.

Ferris, H., R. C. Venette, and K. M. Scow. 2004. Soil management to enhance bacterivore and fungivore nematode populations and their nitrogen mineralization function. Applied Soil Ecology 24:19–35,

Folorunso, O. A., D. E. Rolston, T. Prichard, and D. T. Louie. 1992. Cover crops lower soil surface strength, may improve soil permeability. California Agriculture 46:26–27.

Fonte, S. J., A. Y. Y. Kong, C. van Kessel, P. F. Hendrix, and J. Six. 2006. Influence of earthworm activity on aggregate-associated carbon and nitrogen dynamics differs with agroecosystem management. Soil Biology and Biochemistry 39:1014–1022.

Gale, W. J., C. A. Cambardella, and T. B. Bailey. 2000. Root-derived carbon and the formation and stabilization of aggregates. Soil Science Society of America Journal 64:201–207.

Garcia-Gomez, A., M. P. Bernal, and A. Roig. 2005. Organic matter fractions involved in degradation and humification processes during composting. Compost Science and Utilization 13:127–135.

Ghodrati, M., M. Chendorain, and Y. J. Chang. 1999. Characterization of macropore flow mechanisms in soil by means of a split macropore column. Soil Science Society of America Journal 63:1093–1101.

Hadas, A., T. A. Doane, A. W. Kramer, C. van Kessel, and W. R. Horwath. 2002. Modelling the turnover of ^{15}N-labelled fertilizer and cover crop in soil and its recovery by maize. European Journal of Soil Science 53:541–552.

Harrison, H. E., D. M. Jackson, A. P. Keinath, P. C. Marino, and T. Pullaro. 2004. Broccoli production in cowpea, soybean, and velvetbean cover crop mulches. HortTechnology 14:484–487.

Havlin, J. L., J. D. Beaton, S. L. Tisdale, and W. L. Nelson. 2005. Soil fertility and fertilizers. 7th ed. Upper Saddle River, NJ: Prentice Hall.

Hirth, J. R., B. M. McKenzie, and J. M. Tisdall. 1996. Volume density of earthworm burrows in compacted cores of soil as estimated by direct and indirect methods. Biology and Fertility of Soils 21:171–176.

Horwath, W. R. 2002. Soil microbial biomass. In G. Bitton, ed., Encyclopedia of environmental microbiology. New York: Academic Press. 663–670.

———. 2007. Carbon cycling and formation of soil organic matter. In E. A. Paul, ed., Soil microbiology, ecology and biochemistry. New York: Academic Press. 303–339.

Horwath, W. R., O. C. Devêvre, T. A. Doane, A. W. Kramer, and C. van Kessel. 2002. Soil C sequestration management effects on N cycling and availability. In J. M. Kimble, et al., eds., Agricultural practices and policies for carbon sequestration in soil. Boca Raton, FL: Lewis. 155–164.

Hutchinson, J. B. 1965. Essays on crop plant evolution. Cambridge, UK: Cambridge University Press.

Jastrow, J. D. 1996. Soil aggregate formation and the accrual of particulate and mineral-associated organic matter. Soil Biology and Biochemistry 28:665–676.

Kong, A. Y. Y., J. Six, D. C. Bryant, R. F. Denison, and C. van Kessel. 2005. The relationship between carbon input, aggregation, and soil organic carbon stabilization in sustainable cropping systems. Soil Science Society of America Journal 69:1078–1085.

Kong, A. Y. Y., S. J. Fonte, C. van Kessel, and J. Six. 2007. Soil aggregates control N cycling efficiency in long-term conventional and alternative cropping systems. Nutrient Cycling in Agroecosystems 79:45–58.

Kuschera, L. 1960. Wurzelatlas Mitteleuropäischer Ackerunkräuter and Kulturpflanzen. Frankfurt am Main: D.L.G.-Verlags.

Lal, R., J. M. Kimble, R. F. Follett, and C. V. Cole. 1998. The potential of U.S. cropland to sequester carbon and mitigate the greenhouse effect. Chelsea, MI: Ann Arbor Press.

Lartey, R. T., E. A. Curl, and C. M. Peterson. 1994. Interactions of mycophagous Collembola and biological-control fungi in the suppression of *Rhizoctonia solani*. Soil Biology and Biochemistry 26:81–88.

Lee, K. E. 1987. Peregrine species of earthworms. In A. M. Bonvicini Pagliai and P. Omodeo, eds., On earthworms: Selected symposia and monographs. Modena, Italy: U.Z.I. Mucchi Editore. 315–327.

Lee, K. E., and R. C. Foster. 1991. Soil fauna and soil structure. Australian Journal of Soil Research 29:745–775.

Lee, K. E., and K. R. J. Smettem. 1995. Identification and manipulation of soil biopores for the management of subsoil problems. In N. S. Jayawardane and B. A. Stewart, eds., Advances in soil science. Boca Raton, FL: Lewis. 211–244.

Liu, A. G., B. L. Ma, and A. A. Bomke. 2005. Effects of cover crops on soil aggregate stability, total organic carbon, and polysaccharides. Soil Science Society of America Journal 69:2041–2048.

Luxmoore, R. J. 1981. Micro-, meso-, and macroporosity of soil. Soil Science Society of America Journal 45:671–672.

Luxmoore, R. J., P. M. Jardine, G. V. Wilson, J. R. Jones, and L. W. Zelazny. 1990. Physical and chemical controls of preferred path flow through a forested hillslope. Geoderma 46:139–154.

Maraun, M., J. Alphei, M. Bonkowski, R. Buryn, S. Migge, M. Peter, M. Schaefer, and S. Scheu. 1999. Middens of the earthworm *Lumbricus terrestris* (Lumbricidae): Microhabitats for micro- and mesofauna in forest soil. Pedobiologia 43:276–287.

Martin, J. P., and K. Haider. 1971. Microbial activity in relation to soil humus formation. Soil Science 111:54–63.

McGarry, D., B. J. Bridge, and B. J. Radford. 2000. Contrasting soil physical properties after zero and traditional tillage of an alluvial soil in the semi-arid subtropics. Soil and Tillage Research 53:105–115.

McVay, K. A., D. E. Radcliffe, and W. L. Hargrove. 1989. Winter legume effects on soil properties and nitrogen fertilizer requirements. Soil Science Society of America Journal 53:1856–1862.

Mendes, I. C., A. K. Bandick, R. P. Dick, and P. J. Bottomley. 1999. Microbial biomass and activities in soil aggregates affected by winter cover crops. Soil Science Society of America Journal 63:873–881.

Merwin, I. A., J. A. Ray, T. S. Steenhuis, and J. Boll. 1996. Groundcover management systems influence fungicide and nitrate-N concentrations in leachate and runoff from a New York apple orchard. Journal of the American Society for Horticultural Science 121:249–257.

Miller, M., and R. P. Dick. 1995. Dynamics of soil C and microbial biomass in whole soil and aggregates in 2 cropping systems. Applied Soil Ecology 2:253–261.

Miller, J. J., B. J. Lamond, N. J. Sweetland, and F. J. Larney. 1999. Preferential leaching in large undisturbed soil blocks from conventional tillage and no-till fields in southern Alberta. Water Quality Research Journal of Canada 34:249–266.

Mitchell, J. P., D. W. Peters, and C. Shennan. 1999. Changes in soil water storage in winter fallowed and cover cropped soils. Journal of Sustainable Agriculture 15:19–31.

Moore, J. C., and H. W. Hunt. 1988. Resource compartmentation and the stability of real ecosystems. Nature 333:261–263.

Moore, J. C., E. L. Berlow, D. C. Coleman, P. C. De Ruiter, Q. Dong, et al. 2004. Detritus, trophic dynamics and biodiversity. Ecology Letters 7:584–600.

Oades, J. M. 1984. Soil organic matter and structural stability: Mechanisms and implications for management. Plant and Soil 76:319–337.

Paul, E. A. 1984. Dynamics of organic matter in soils. Plant and Soil 76:275–285.

Perret, J., S. O. Prasher, A. Kantzas, and C. Langford. 1999. Dimensional quantification of macropore networks in undisturbed soil cores. Soil Science Society of America Journal 63:1530–1543.

Pitkanen, J., and V. Nuutinen. 1997. Distribution and abundance of burrows formed by *Lumbricus terrestris* L and *Aporrectodea caliginosa* Sav in the soil profile. Soil Biology and Biochemistry 29:463–467.

Poudel, D. D., W. R. Horwath, J. P. Mitchell, and S. R. Temple. 2001. Impacts of cropping systems on soil nitrogen storage and loss. Agricultural Systems 68:253–268.

Puget, P., E. Besnard, and C. Chenu. 1996. Particulate organic matter fractionation according to its location in soil aggregate structure. Comptes Rendus de l'Academie des Sciences [Paris], Serie II 322:965–972.

Quemada, M., and M. L. Cabrera. 1995. Carbon and nitrogen mineralized from leaves and stems of four cover crops. Soil Science Society of America Journal 59:471–477.

Ranells, N. N., and M. G. Wagger. 1992. Nitrogen release from crimson clover in relation to plant growth stage and composition. Agronomy Journal 84:424–430.

———. 1997a. Grass-legume bicultures as winter annual cover crops. Agronomy Journal 89:659–665.

———. 1997b. Winter annual grass-legume bicultures for efficient nitrogen management in no-till corn. Agriculture Ecosystems and Environment 65:23–32.

Roberson, E. B., S. Sarig, and M. K. Firestone. 1991. Cover crop management of polysaccharide-mediated aggregation in an orchard soil. Soil Science Society of America Journal 55:734–739.

Roseberg, R. J., and E. L. McCoy. 1992. Tillage-induced and traffic-induced changes in macroporosity and macropore continuity–air permeability assessment. Soil Science Society of America Journal 56:1261–1267.

Scheu, S. 2003. Effects of earthworms on plant growth: Patterns and perspectives. Pedobiologia 47:846–856.

Scheu, S., A. Theenhaus, and T. H. Jones. 1999. Links between the detritivore and the herbivore system: Effects of earthworms and Collembola on plant growth and aphid development. Oecologia 119:541–551.

Schjønning, P., S. Elmholt, L. J. Munkholm, and K. Debosz. 2002. Soil quality aspects of humid sandy loams as influence by organic and conventional long-term management. Agriculture, Ecosystems and Environment 88:195–214.

Schrader, S., M. Joschko, H. Kula, and O. Larink. 1995. Earthworm effects on soil structure with emphasis on soil stability and soil water movement. In K. H. Hartge and B. A. Stewart, eds., Soil structure, its development and function. Boca Raton, FL: CRC Press. 109–133.

Schutter, M. E., J. M. Sandeno, and R. P. Dick. 2001. Seasonal, soil type, and alternative management influences on microbial communities of vegetable cropping systems. Biology and Fertility of Soils 34:397–410.

Shuster, W. D., S. Subler, and E. L. McCoy. 2000. Foraging by deep-burrowing earthworms degrades surface soil structure of a fluventic Hapludoll in Ohio. Soil and Tillage Research 54:179–189.

Six, J., E. T. Elliott, and K. Paustian. 2000. Soil macroaggregate turnover and microaggregate formation: A mechanism for C sequestration under no-tillage agriculture. Soil Biology and Biochemistry 32:2099–2103.

Six, J., H. Bossuyt, S. Degryze, and K. Denef. 2004. A history of research on the link between (micro)aggregates, soil biota, and soil organic matter dynamics. Soil Tillage and Research 79:7–31.

Smettem, K. R. J., and N. Collis-George. 1985. The influence of cylindrical macropores on steady-state infiltration in a soil under pasture. Journal of Hydrology 79:107–114.

Smith, M. S., W. W. Frye, and J. J. Varco. 1987. Legume winter cover crops. Advances in Soil Science 7:95–139.

Soane, B. D. 1988. The role of organic matter in soil compaction. Journal of the Science of Food and Agriculture 45:134–135.

Springett, J., and R. Gray. 1997. The interaction between plant roots and earthworm burrows in pasture. Soil Biology and Biochemistry 29:621–625.

Springett, J., R. A. J. Gray, and J. B. Reid. 1992. Effect of introducing earthworms into horticultural land previously denuded of earthworms. Soil Biology and Biochemistry 24:1615–1622.

Stevenson, F. J. 1994. Humus chemistry: Genesis, composition, reactions. New York: Wiley.

Stirzaker, R. J., and I. White. 1995. Amelioration of soil compaction by a cover-crop for no-tillage lettuce production. Australian Journal of Agricultural Research 46:553–568.

Stivers, L. J., and C. Shennan. 1991. Meeting the nitrogen needs of processing tomatoes through winter cover cropping. Journal of Production Agriculture 4:330–335.

Terman, G. L. 1979. Volatilization losses of nitrogen as ammonia from surface-applied fertilizers: Organic amendments and crop residues. Advances in Agronomy 31:189–223.

Thompson, M. L., H. Zhang, M. Kazemi, and J. A. Sandor. 1989. Contribution of organic matter to cation exchange capacity and specific surface area of fractionated soil materials. Soil Science 148:250–257.

Tisdale, S. L., and W. L. Nelson. 1975. Soil fertility past and present. In S. L. Tisdale and W. L. Nelson, eds., Soil fertility and fertilizers. 3rd ed. New York: Macmillan. 5–20.

Tisdall, J. M., and J. M. Oades. 1982. Organic matter and water-stable aggregates in soils. Journal of Soil Science 62:141–163.

Veenstra, J. J., J. P. Mitchell, and W. R. Horwath. 2008. Conservation tillage and cover cropping improve soil properties in the San Joaquin Valley. California Agriculture 60:146–153.

Vos, J., P. E. L. van der Putten, M. H. Hussein, A. M. van Dam, and P. A. Leffelaar. 1998. Field observations on nitrogen catch crops. II. Root length and root length distribution in relation to species and nitrogen supply. Plant and Soil 20:149–155.

Whalley, W. R., E. Dumitru, and A. R. Dexter. 1995. Biological effects of soil compaction. Soil and Tillage Research 35:53–68.

Williams, W. A. 1966. Management of nonleguminous green manures and crop residues to improve the infiltration rate of an irrigated soil. Soil Science Society of America Proceedings 30:631–634.

Willoughby, G. L., E. J. Kladivko, and M. R. Savabi. 1997. Seasonal variations in infiltration rate under no-till and conventional (disk) tillage systems as affected by *Lumbricus terrestris* activity. Soil Biology and Biochemistry 29:481–484.

Zhang, H., and S. Schrader. 1993. Earthworm effects on selected physical and chemical properties of soil aggregates. Biology and Fertility of Soils 15:229–234.

Zhu, Y. G., Y. Q. He, S. E. Smith, and F. A. Smith. 2002. Buckwheat (*Fagopyrum esculentum* Moench) has high capacity to take up phosphorus (P) from a calcium (Ca)-bound source. Plant and Soil 239:1–8.

4 *Michael D. Cahn and Gene Miyao*

Water Management and Impacts on Water Quality

Water Use by Cover Crops

Factors Affecting Water Use

Cover crops require water for germination (stand establishment) and for growth of the canopy and root system. Supplemental irrigation may be needed, depending on when and where the crop is planted in California. In high-rainfall regions of the state (> 24 inches per year), rain often begins early enough to germinate crops planted during late fall and early winter. However, irrigation may still be needed for germination if rains are untimely, especially for slow-growing crops such as legumes. Rainfall can usually sustain the growth of cover crops during the late fall and winter because evapotranspiration requirements are commonly lower than the total precipitation. In dry areas of the state, irrigation is often needed for germination, especially if the growing period is limited or early fall establishment is desired. Cover crops seeded from spring through early fall usually require irrigation for establishing the stand and to replace water lost by evapotranspiration.

Water use by a cover crop depends on the total number of days the crop is grown, the time of year planted, and the canopy characteristics of the species mix. Because of the longer days and warmer temperatures, the estimated water use of a rye cover crop planted on the Central Coast in the summer is more than twice that of the crop planted there in the late fall. A study in the Salinas Valley showed that legume-rye mixes extract less water than rye or mustard during the first 2 months of growth, presumably because the legume species grew slower than the rye and mustard (Brennan and Cahn, unpublished) (fig. 4.1A). Also, seeding rate was found to affect water use. Water use was highest for mixes planted at 3 times the normal seeding rates due to greater canopy cover early in the season (fig. 4.1B). Additionally, root growth and water extraction was found to be deeper for the rye and mustard crops than in the the legume-rye mixture.

Deficit moisture conditions, which can occur in nonirrigated cover crops as the soil dries, reduce growth and water use. Low soil moisture may induce flowering or cause early senesce of foliage.

Estimating Water Use

Estimating the water consumption of cover crops can help optimize their management in vegetable crop systems, including deciding when to seed and terminate the cover crop, how much water to budget, and when to schedule irrigations. Calculations of the daily evapotranspiration can be used to estimate the water consumption of cover crops. Evapotranspiration (ET) is the combination of water lost by evaporation from the soil surface and by transpiration from the leaves of the crop. ET calculations are typically expressed in units of inches per day or millimeters per day.

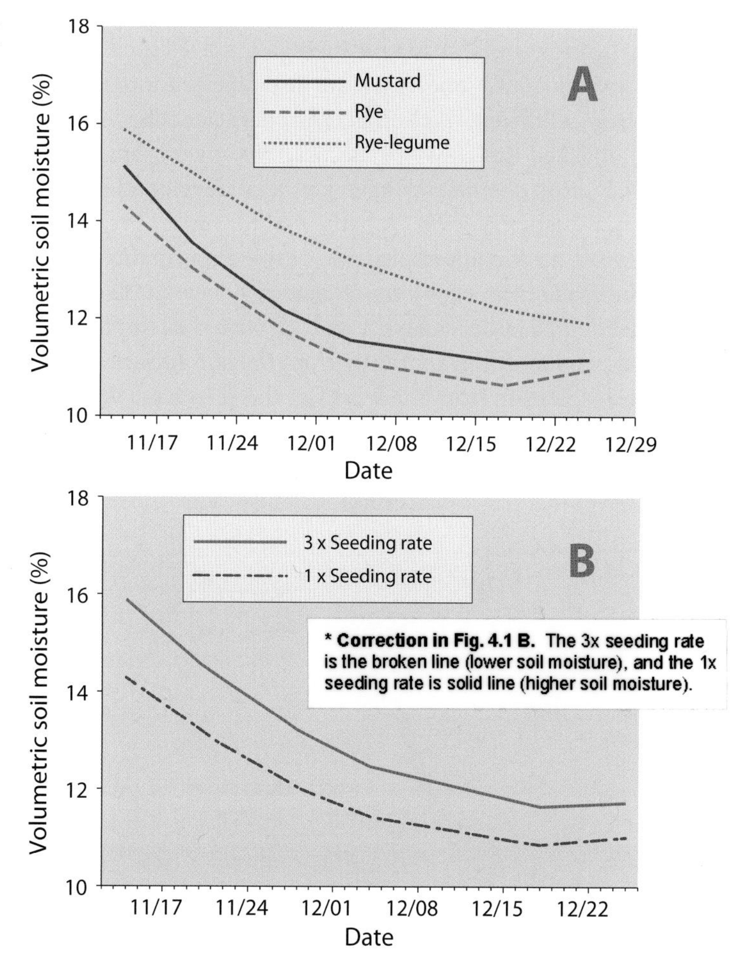

Figure 4.1. Effects of cover crop mixes (A) and the seeding rate of the mixes (B) on the soil moisture content of the top 18 inches of a sandy loam soil.

Table 4.1. Crop coefficients for selected cover crop species and growth stages

Crop	Crop coefficients (K_c)		
	Stand establishment	Full cover	Maturity
grazed or mowed pasture	—	0.90	—
barley*	0.23	1.04	0.01
grass–clover pasture	—	1.05	—
wheat†	0.38	1.07	0.15
small grains‡	0.25	1.20	0.40

Notes:
*Imperial Valley, 11/30 planting date.
†Imperial Valley, 12/30 planting date.
‡San Joaquin Valley, 11/01 planting date.

ET requirements of cover crops can be estimated by the following formula using reference ET obtained from the California Irrigation Management and Information System (CIMIS, operated by the California Department of Water Resources) and the appropriate crop coefficient (K_c):

$$ET_{crop} = ET_{ref} \times K_c$$

Daily and historical reference ET data are available from the CIMIS Web site for most agricultural production regions of the state. The crop coefficient usually ranges from 0.1 to 1.1 and is closely related to the percentage of the ground shaded by the canopy. Irrigation method and physiological stages such as flowering and senescence are also factored into the crop coefficient. Although crop coefficients have been published for a few species used for cover crops (table 4.1), crop coefficients have not been developed for many other species. Nonetheless, for most cover crop mixes a good approximation of the crop coefficient can be made from estimates of canopy cover. At planting, a coefficient between 0.1 and 0.3 would be appropriate to account for evaporation from the soil surface and transpiration from the leaves of the seedlings. When the crop reaches 10% canopy cover, the coefficient would increase proportionally to the percentage of canopy cover (fig. 4.2). At maximum canopy size, the crop coefficient may range from 0.9 to 1.1, depending on the characteristics of the species mix. As the crop reaches maturity and leaves begin to senesce, the coefficient may decrease to 80% of the maximum value before termination and incorporation of the crop (see fig. 4.2). If the canopy dies back completely, a coefficient of less than 0.1 would be appropriate, since transpiration is not occurring.

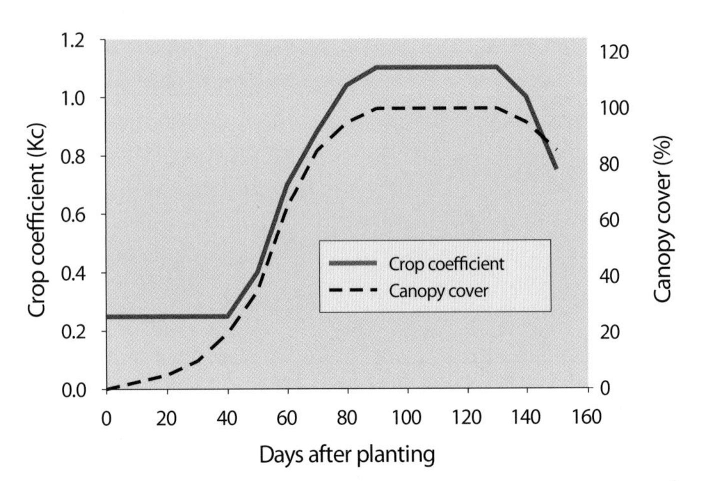

Figure 4.2. Canopy cover and crop coefficient values estimated for a rye cover crop planted in early fall.

ET requirements of cover crops can be substantial during the summer months. In inland regions of the state such as Five Points in Fresno County, ET requirements could be as high as 0.3 inches per day during late June through early August, while coastal areas subject to fog, such as Watsonville in Santa Cruz County, could be as low as 0.1 inches per day during the same period.

Water Management of Vegetable Crops after Cover Cropping

Cover crops may increase the amount of water available for subsequent vegetable crops by improving soil structure. Root exudates and organic matter contributed by the decomposition of a cover crop can increase soil tilth, macropore structure, and water-holding capacity (see chapter 3). Enhancing soil structure can increase infiltration, minimizing irrigation runoff from overhead sprinkler and furrow systems. However, several cycles of cover crops may be needed before changes in soil physical properties become noticeable. Aggregation of soil is closely linked to soil organic matter content, which usually changes slowly as cover crops are incorporated into the production system. Increasing aggregation and macropore structure can encourage deep rooting of vegetable crops and can increase the effective depth of water extraction. Usually, a deep-rooted plant can tolerate less-frequent watering than can a shallow-rooted plant. Improving the aggregate structure of the soil may also affect lateral movement of moisture from drip irrigation lines (as well as vertical movement); however, whether movement would be enhanced or diminished may depend on texture and other characteristics of the soil. Increasing the size or number of soil macropores can increase infiltration but can also reduce the ability to move water to the tail end of furrow-irrigated fields, especially in nontrafficked furrows (e.g., guess rows). To compensate for higher infiltration rates and longer irrigation sets, the management of furrow irrigation may need to be adjusted. Methods such as "torpedoing" furrow bottoms or using surge irrigation can increase the rate that water advances down the furrows during irrigations.

Water extraction by winter cover crops often affects soil moisture conditions when fields are being prepared for a subsequent vegetable crop. Ideally, termination of the cover crop will be timed to optimize soil moisture for tillage and planting operations. Fall-planted cover crops begin to have high water-use requirements in the early spring, when they reach maximum canopy size and warmer weather and longer days increase consumptive water use. Cover crops at this stage of growth can help dry out saturated soils so that tillage operations can be started as early as possible. However, late termination of a cover crop can dry soil so much that irrigation may be needed before tillage operations or for germinating the vegetable crop. Close monitoring of soil moisture and weather conditions during the late stages of a cover crop can help growers time the termination of the cover crop so that soil moisture is optimal for tillage and planting.

Another problem observed in maturing cover crops is that the soil can become wet on the surface but remain dry deeper in the profile. The combination of spring showers and shade from the mature cover crop, which reduces evaporation, keeps the soil surface wet, but extraction of moisture by deep roots may dry the soil below the wet surface layer. The different moisture contents at the soil surface and deeper down makes incorporating the cover crops difficult to time. Delaying incorporation of a cover crop until the moisture at the soil surface is ideal could sacrifice moisture needed deeper in the soil profile to sustain the subsequent vegetable crop and to facilitate preplant tillage operations such as bed shaping.

In regions of the state where soils are affected by salts, maximizing infiltration of rainfall to leach salt near the soil surface can maintain soil productivity. Cover crops can help with salt management by enhancing infiltration through improvements in soil structure, but cover crops also consume water, which could reduce drainage and leaching during winter rain events. In areas where salinity management is critical for vegetable crop production, cover crops should be terminated before water consumption is high to maximize leaching of salts by rain.

Erosion and Water Quality Benefits

Winter-planted cover crops are increasingly grown to reduce storm water runoff from agricultural lands and to prevent the migration of sediment, nutrients, and pesticides to surface water bodies. Cover crops may also trap mobile nutrients such as nitrate before they can leach into groundwater. A study on the Central Coast reported that winter cover crops of mustards, phacelia, and rye all significantly reduced soil nitrate levels compared with bare plots (Jackson et al. 1993). On average these species accumulated 143 pounds of nitrogen per acre in the above- and belowground biomass during a 17-week period. Oilseed radish and white radish accumulated the most nitrogen (181 lb/ac N), and Merced rye and annual ryegrass accumulated the least (96 lb/ac N).

Cover crops reduce winter runoff by maintaining high infiltration rates into soil. Initial infiltration rates of soil can be 10 to 20 times higher under unsaturated conditions than under saturated conditions. Storm events usually produce significantly more runoff in fields that are initially wet rather than dry. By extracting soil moisture, cover crops allow a greater portion of rainfall to infiltrate, reducing the portion available for runoff. Cover crops may also enhance infiltration by increasing the macropore volume of the soil. Root exudates and other organic material from the cover crop help aggregate soil particles and contribute to the large pores that allow water to rapidly infiltrate into the soil profile. By intercepting rainfall, cover crops absorb energy from the raindrops that can break down aggregates on the soil surface and cause the formation of a crust, which can greatly impede infiltration and increase runoff.

Cover crops may also reduce suspended sediments and turbidity of storm water runoff by slowing the movement of runoff. Fast-flowing water has more energy to detach and lift soil particles and keep sediments suspended than does slow-flowing water. Also, slow runoff has more time to infiltrate into soil than faster runoff as water recedes from a field.

Cover crops have limited ability to increase infiltration if physical barriers such as hardpans and clay layers, which limit drainage, are present in the subsoil. Shallow water tables may also limit drainage. Deep tillage, addition of soil amendments, or in some cases tile drains may be required to maximize the water quality benefits of cover crops. In a study in which cover crops were planted in furrows between strawberry beds and a clay layer was present a few feet

Table 4.2. Effect of cover crop treatment on turbidity and amount of runoff in rain-simulated trials using sprinklers

Cover crop treatment	Turbidity (NTU*)	Runoff (% of rainfall†)
bare soil	770	4.5
triticale only in furrows	359	1.5
rye on listed beds and in furrow	47	0.2

Notes:
*Low NTU (nephelometric turbidity units) indicates less turbidity.
†Rainfall simulated using sprinklers.

below the soil surface, storm runoff was not reduced compared with bare (unplanted) furrows (Cahn et al. 2006a). However, sediment concentration and turbidity of the runoff collected from the cover cropped furrows were reduced to just above half the values measured in the bare furrows. In another trial in which the field was well drained, storm runoff and turbidity were significantly reduced by rye broadcast-planted on beds and furrows and by triticale planted only in furrows compared with bare plots (Cahn et al. 2006b) (table 4.2).

Nutrients (nitrogen and phosphorus) and pesticides attached to suspended soil particles in runoff and carried into surface water represent important sources of impairments to water quality. By reducing sediment concentration in the runoff, cover crops can reduce total nitrogen and phosphorus concentrations in the runoff. However, concentrations of soluble nutrients (nitrate and orthophosphate) are usually not lowered by reducing sediment concentration. Nevertheless, cover crops are effective in reducing the load of soluble nutrients by lessening the total volume of runoff. The situation is similar with pesticides. By reducing sediment concentrations, cover crops reduce the movement of sediment-bound pesticides. While cover crops may not reduce the concentration of soluble pesticides in runoff, by reducing runoff volume the amount of pesticide leaving the field is reduced.

Strategies for growing cover crops to control storm water runoff and protect water quality are being refined to fit existing production constraints or for use with other practices to maximize water quality benefits. For example, in the Sacramento Valley winter cover crops are often planted only on tops of beds prepared for processing tomatoes to prevent regrowth from the furrow bottoms (fig. 4.3). This strategy also allows the furrow bottoms to dry out rapidly in the spring and facilitate tractor operations when planting

Figure 4.3. Wheat *(Triticum aestivum)* (left) and triticale (*Triticale hexaploide* Lart.) (right) cover crops planted on 60-inch-wide beds used for production of processing tomatoes. E. M. Miyao

Figure 4.4. Triticale (*Triticale hexaploide* Lart.) planted in furrow bottoms of listed beds (left) and Merced Rye (*Secale cereale* L.) planted in furrows and on tops of listed beds (right). M. D. CAHN

begins. Although there is no cover crop in the furrows to slow runoff during storm events, the cover crop planted on the beds can increase infiltration of storm water and trap potentially leachable nitrogen. Disking the unplanted furrows in the fall can be an alternative method to slow runoff. On the Central Coast, where beds are often listed in the fall but not shaped until spring, the period between cover crop incorporation and planting of the vegetable crop can be hastened by broadcast-seeding a cover crop on the beds rather than on flat ground. Also, low-stature cover crops such as phacelia or triticale (fig. 4.4) planted in the furrow bottoms can reduce runoff without interfering with spring tillage and planting operations. Additionally, zones such as the seedline on beds can be left unplanted so the cover crop residue is less likely to clog tillage and seeding implements.

Conservation tillage methods can also lessen the difficulties of incorporating the cover crop in the spring and enhance water quality benefits. A winter cover crop planted on beds can be chemically or mechanically terminated and chopped into mulch with a flail mower. Using no-till transplanting equipment fitted with coulters to cut through residue on the seedline, transplants can be planted directly into the mulch. The residue on the shoulders of the bed can be incorporated after planting or left to decompose on the soil surface. The reduced tillage can improve macropore structure by accelerating the buildup of organic matter. Choosing fast-decomposing species such as mustards over slow-decomposing species such as barley can also reduce residue management challenges for the subsequent vegetable crop.

References

Cahn, M., M. Bolda, and R. Smith. 2006a. Winter cover crops for reducing storm runoff and protecting water quality in strawberries. Monterey County Crop Notes (Nov.-Dec.): 2-4. UCCE Monterey County Web site, http://cemonterey.ucdavis.edu.

Cahn, M. D., R. F. Smith, and A. Young. 2006b. Evaluation of practices for controlling storm run off from vegetable fields. Monterey County Crop Notes (Nov./Dec.). UCCE Monterey County Web site, http://cemonterey.ucdavis.edu/newsletterfiles/Monterey_County_Crop_Notes10453.pdf.

Dabney, S. M. 1998. Cover crop impacts on watershed hydrology. Soil and Water Conservation 53(3): 207–213.

Dabney, S. M., J. A. Delgado, and D. W. Reeves. 2001. Using winter cover crops to improve soil and water quality. Communications in Soil Science and Plant Analysis 32(7–8): 1221–1250.

Jackson, L. E. 2000. Fate and losses of nitrogen from a nitrogen-15 labeled cover crop in an intensively managed vegetable system. Soil Science Society of America Journal 64:1404–1412.

Jackson, L. E., L. J. Wyman, and L. J. Stivers. 1993. Winter cover crops to minimize nitrate losses in intensive lettuce production. Journal of Agricultural Science 121:55–62.

Goldhammer, D. A., and R. L. Snyder. 1989. Irrigation scheduling: A guide for efficient on-farm water management. Oakland: University of California Division of Agriculture and Natural Resources Publication 21454.

Madden, N. M., J. P. Mitchell, W. T. Lanini, M. D. Cahn, et al. 2004. Evaluation of conservation tillage and cover crop systems for organic processing tomato production. HortTechnology 14(2): 243–250.

Wyland, L. J., L. E. Jackson, and K. F. Schulbach. 1995. Soil-plant nitrogen dynamics following incorporation of a mature rye cover crop in a lettuce production system. Journal of Agricultural Science 124:17–25.

Wyland, L. J., L. E. Jackson, W. E. Chaney, K. Klonskey, S. T. Koike, and B. Kimple. 1996. Winter cover crops in a vegetable cropping system: Impacts on nitrate leaching, soil water, crop yield, pests, and management costs. Agricultural Ecosystems and Environment 59:1–17.

5 Soil Nitrogen Fertility Management

Mark Gaskell, Richard Smith, Louise E. Jackson, and Timothy K. Hartz

Cover crops can play an important role in managing nitrogen (N) in vegetable production systems. Vegetable production generally requires high amounts of readily available nitrogen, but at the end of the production season, the soil can contain high levels of residual soil nitrate-nitrogen. Winter cover crops absorb residual soil nitrate, reducing nitrate leaching and denitrification, a process that causes soil emissions of nitrous oxide, an important greenhouse gas. All cover crops absorb soil nitrate, but leguminous cover crops can also convert atmospheric nitrogen to plant-available forms and enrich the soil with nitrogen as they decompose. The choice of cover crop for nitrogen management depends largely on the production system and the intended goal of the cover crop. Nonleguminous cover crops such as grasses and mustards are preferred in situations where residual soil nitrate is high in the fall. Legumes and grass-legume mix cover crops are used to build up nitrogen levels in the soil for succeeding vegetable crops, particularly in organic production.

Nitrogen Trapping

Nitrate, the most common mineral form of soil nitrogen, is easily moved downward in the soil profile by leaching with water. Large amounts of nitrate-nitrogen may remain in the soil at the end of the vegetable growing season in the fall, and this nitrate is at risk of being leached with winter rains or converted to nitrous oxide and lost as atmospheric nitrogen (N_2). Winter cover crops provide an important cultural practice for capturing residual nitrate in vegetable production fields and sequestering it in plant biomass during the rainy period, thereby reducing nitrate-nitrogen losses. Studies in the California coastal vegetable production region have shown that Merced rye can reduce leaching of nitrate by 70% (Jackson et al. 1993). Winter-grown grass and mustard cover crops typically absorb 100 to 200 pounds of nitrogen per acre in their aboveground biomass, capturing or trapping nitrogen that would otherwise be lost and contribute to environmental contamination. Legumes can also scavenge residual nitrogen from soil, but grasses and mustards are more efficient at this process. Mixes of grasses and legumes, however, blend species that are efficient at both nitrogen trapping and nitrogen fixation.

Nitrogen Fixation

Legumes are unique because they convert, or "fix," atmospheric nitrogen, which is not otherwise available to plants, to forms of nitrogen that plants can use. These plants increase the amount of nitrogen in the soil available to the succeeding crop. Growing legume cover crops is important in organic production, as they provide an economical source of nitrogen nutrition (see chapter 12). Legumes used as cover crops in vegetable production typically contain 100 to 200 pounds of nitrogen per acre. However, fixation by legumes is suppressed by high levels of mineral nitrogen (i.e., nitrate and ammonium) in the soil. In sites with high nitrogen fertility, legumes use the readily available source of soil nitrogen instead of fixing atmospheric nitrogen. See chapter 2 for more discussion of nitrogen fixation by legumes.

Nitrogen for Crop Growth after Cover Crop Incorporation

Regardless of whether it was trapped by nonlegumes or fixed by legumes, some of the nitrogen contained in cover crop biomass will be released as the cover crop decomposes after incorporation into moist soil in the spring. The process of converting the organic nitrogen contained in the cover crop to mineral nitrogen (mineralization) is carried out by soil micro- and macrofauna (e.g., protozoa, nematodes, soil insects, and worms) and microbes (bacteria and fungi). Mineralization depends on many factors, but it is most rapid in warm (> 60°F), moist soil where soil microbes are abundant and active. In addition, the species of cover crop and the quantity of nitrogen relative to carbon (C) in the residue affects the speed of mineralization.

The rate of nitrogen mineralization from incorporated cover crops is more rapid in legumes than in nonlegumes because legumes contain a higher level of nitrogen (lower carbon-to-nitrogen, or C:N, ratio), which facilitates nitrogen mineralization. Mature grass cover crops have lower levels of nitrogen and higher carbon in their biomass, and as a result, nitrogen contained in their biomass is mineralized more slowly and less is available for subsequent vegetable crop growth.

The reason for the small initial nitrogen release from cover crop residue with a high C:N ratio is that carbon stimulates the growth of soil microbes that, to meet their own nitrogen demand, rapidly assimilate the cover crop nitrogen as well as the soil nitrogen. This process immobilizes nitrogen, tying it up so that it is less available to plants, at least until the soil microorganisms begin to die or are consumed by other organisms so that their nitrogen converts to ammonium and nitrate (see table 5.1). It is best to allow extra time between incorporating a mature grass cover crop and planting a vegetable crop to prevent the early growth of the vegetable crop from being affected by nitrogen immobilization. To speed decomposition of grass residues with high levels of carbon, small amounts of nitrogen fertilizer (e.g., 20 to 30 lb/ac N) may be added to offset the immobilization by soil microorganisms.

The rate of nitrogen release from mustard cover crops is intermediate between legumes and mature grasses. Legume-grass cover crop mixes also have intermediate nitrogen release rates, depending on the characteristics and proportions of the species.

In general, cover cropping as a source of nitrogen is a more important component of organic than conventional production systems. In organic systems, nitrogen from cover crops is an important and cost-effective part of the nitrogen fertilization regime for certain short-season crops. From 4 to 30% of the nitrogen contained in cover crops is directly used by subsequent vegetable crops (Jackson 2000), but the overall fertilizer nitrogen replacement value of a vigorous legume cover crop can be in excess of 100 pounds of nitrogen per acre. In a recent study comparing organic bell peppers grown with and without a legume mix cover crop (90% legume and 10% barley) with varying rates of nitrogen, the peppers without the prior cover crop required 200 pounds of nitrogen per acre to reach yields similar to peppers with a prior cover crop that received only 100 pounds of nitrogen per acre (Gaskell 2004). In addition, over a 14-week sampling period, the weekly soil nitrate-nitrogen level of the cover crop plus fertilizer treatments exceeded that of fertilizer-only plots even at fertilizer rates of 200 pounds of nitrogen per acre.

Nitrogen release from cover crops continues beyond the initial rapid decomposition period, but at a reduced rate. Microbes and other soil biota that flourish when the plant residue was initially incorporated gradually die or are consumed by their predators so that their nitrogen is taken up by other soil biota or by plants. Thus, cover crops generate a more active soil food web and extend the period of nitrogen release. However, the majority of nitrogen mineralization from a cover crop occurs during the first 3 to 8 weeks after incorporation into the soil.

Synchrony of Nitrogen Availability

The release of nitrogen from cover crops follows the general pattern shown in figure 5.1. There is a flush of nitrogen release followed by a return to background

Table 5.1. Effect of cover crop nitrogen content on nitrogen release

Percentage of N in cover crop	Effect of N release	Examples of cover crops
0.5	will tie up N	cereal straw
1.0	will tie up N	cereal straw
1.5	will tie up N	cereal at heading
2.0	may tie up N*	cereal before heading
2.5	may tie up N*	mustards at heading and immature cereal
3.0	will release N	mustards, legumes, and juvenile cereal
3.5	will release N	legumes and immature mustards
4.0	will release N	legumes

Note: *Depends upon levels of mineral N in the soil.

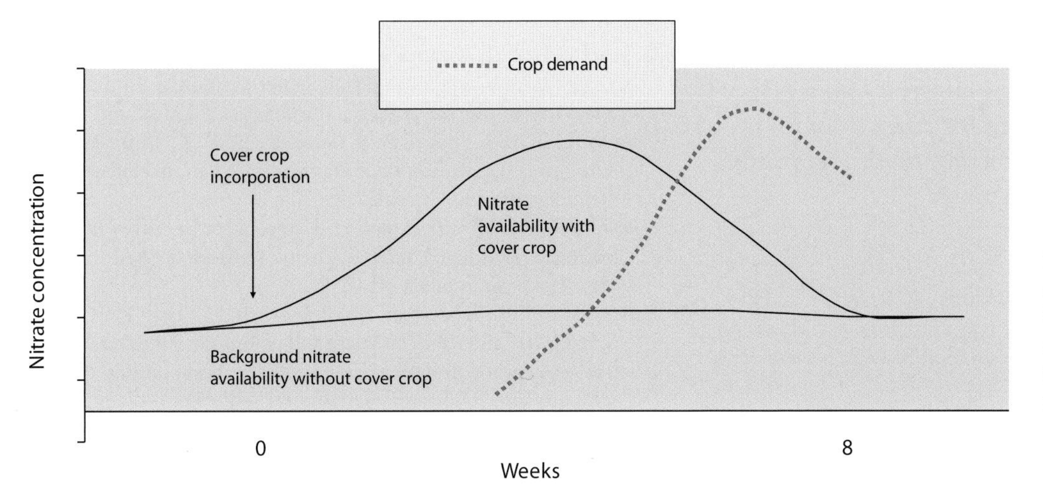

Figure 5.1. Typical pattern of nitrate availability from a legume cover crop assuming no leaching rain or irrigation. *Source:* Adapted from Gaskell and Smith 2007

levels within 8 weeks or so. This general pattern may not fully explain the dynamics that are at work upon incorporation of cover crop residue: the pattern does not illustrate the partitioning of nitrogen between the subsequent crop and the nitrogen immobilized by microbial activity. For example, in a study evaluating soil nitrogen dynamics after incorporation of a phacelia cover crop, soil nitrate-nitrogen increased over a 10-day period, remained at elevated levels for the following 2-week period, then declined to soil preincorporation levels by harvest of the subsequent lettuce crop (116 days) (Jackson 2000). During the 116-day period, cover crop nitrogen was subject to microbial nitrogen uptake and release, nitrate leaching, and denitrification, but overall, about 10% of the nitrogen contained in the lettuce at harvest was derived from the previous cover crop.

To successfully take advantage of the nitrogen released from a cover crop, the release of mineral nitrogen from soil food web activity should coincide with nitrogen demand by the subsequent vegetable (see fig. 5.1). The dynamics of the turnover and cycling of nitrogen after residue incorporation are complex and variable; further, irrigation or rainfall prior to planting vegetable crops can increase the loss of cover crop-derived nitrogen through leaching. When managed carefully, the nitrogen available from a cover crop may be sufficient for the nitrogen needs of a relatively short-cycle leafy crop such as spinach or leaf lettuce. For most vegetable crops that require a growth period of more than 60 days, however, the amount of nitrogen released from cover crops is not sufficient to meet peak nitrogen demand, and supplemental nitrogen applications are generally required to achieve optimal yield and quality.

For organic growers interested in using cover crop nitrogen as a primary source of nitrogen for the succeeding vegetable crop, incorporating the cover crop should be synchronized with planting the succeeding crop. Mineralization of cover crop-derived nitrogen begins immediately after incorporation into moist soil, so timely soil preparation and seeding or transplanting is essential for the efficient use of available nitrogen. Delays in planting after cover crop incorporation can reduce the amount of nitrogen available to the succeeding crop, or rainfall or preirrigation may leach mineralized nitrate from the cover crop.

Long-Term Impacts on Nitrogen Availability

Cover crops also have long-term effects on nitrogen availability for subsequent crops. Cover crops rich in carbon, such as grasses, provide the building blocks of soil organic matter. Soil organic matter is a reservoir for nitrogen, and the nitrogen contained in organic matter decomposes over a period of years, contributing to the inherent fertility of the soil. Studies of vegetable production systems comparing high-carbon inputs (cover crops and compost) with low-carbon inputs indicate that organic matter and nitrogen storage increases with high-carbon inputs (Poudel et al. 2001; table 5.2). Over a period of years, cover crops can increase the ability of the soil to supply nitrogen to cash crops, and the soil becomes more fertile if organic

Table 5.2. Increase in soil organic matter and soil organic nitrogen after 12 years of organic management (with annual cover cropping and compost use) compared with conventional management (no cover crops or compost)

Cropping system	Soil organic matter (%)	Nitrogen storage in soil organic matter (lb/ac)
short rotation conventional (low carbon input)	1.10	0
organic (high carbon input)	1.46	901

Source: Adapted from Poudel et al. 2001.

matter increases. Carbon-rich residues from grass cover crops provide more long-lasting effects in this regard than do residues from legumes. For cover crops to have long-lasting effects on soil they must be regularly included in vegetable crop rotations.

References

Aulakh, M. S., J. W. Doran, D. T. Walters., A. R. Moier, and D. D. Francis. 1991. Crop residue type and placement effects on denitrification and mineralization. Soil Science Society of America Journal 55:1020–1025.

Creamer, N. G., M. A. Bennett, and B. R. Stinner. 1997. Evaluation of cover crop mixtures for use in vegetable production systems. HortScience 32:866–870.

Ebelhar, S. A., W. W. Frye, and R. L. Blevins. 1984. Nitrogen from legume cover crops for no-tillage corn. Agronomy Journal 76:51–55.

Gaskell, M. 2004. Nitrogen availability, supply and sources in organic row crops. California Organic Production and Farming in the New Millennium: A California Research Symposium, July 15, 2004, Berkeley, CA. UC Davis Sustainable Agriculture Research and Education Program.

Gaskell, M., and R. Smith. 2007. Nitrogen sources for organic vegetable crops. HortScience 17(4): 431–441.

Jackson, L. E. 2000. Fates and losses of nitrogen from a nitrogen-15-labeled cover crop in an intensively managed vegetable system. Soil Science Society of America Journal 64:1404–1412.

Jackson, L. E., L. J. Wyland, and L. J. Stivers. 1993. Winter cover crops to minimize nitrate losses in intensive lettuce production. Journal of Agricultural Science 121:55–62.

Jackson, L. E., L. J. Stivers, B. T. Warden, and K .K. Tanji. 1994. Crop nitrogen utilization and soil nitrate loss in a lettuce field. Fertilizer Research 37:93–105.

Jong-Ho, S., J. J. Meisinger, and H.-J. Lee. 2006. Recovery of nitrogen-15-labeled hairy vetch and fertilizer applied to corn. Agronomy Journal 98:245–254.

Lundquist E. J., L. E. Jackson, K. M. Scow, and C. Hsu. 1991. Changes in microbial biomass and community structure, and soil carbon and nitrogen pools after incorporation of rye into three California agricultural soils. Soil Biology and Biochemistry 31:221–236.

Poudel, D. D., W. R. Horwath, J. P. Mitchell, and S. R. Temple. 2001. Impacts of cropping systems on soil nitrogen storage and loss. Agricultural Systems 68(3): 253–268.

Ranells, N., and M. G. Wagger. 1997. Nitrogen-15 recovery and release by rye and crimson clover cover crops. Soil Science Society of America Journal 61:943–948.

Shennan, C. 1992. Cover crops, nitrogen cycling, and soil properties in semi-irrigated vegetable production systems. HortScience 27:749–754.

Wyland L. J., L. E. Jackson, and K. F. Schulbach. 1995. Soil–plant nitrogen dynamics following incorporation of a mature rye cover crop in a lettuce production system. Journal of Agricultural Science [Cambridge] 124:17–25.

PART 3

Effects of Cover Cropping on Pest Management

6 *Eric B. Brennan, Oleg Daugovish, Steven Fennimore, and Richard Smith*

Weeds

Weeds are often the most common and costly pests in vegetable production. Weeds that germinate during cover cropping and produce seed increase the weed seed bank and may increase production costs. The effects of cover crops on weeds are relatively easy to monitor. For example, the size of individual weeds and weed biomass during cover cropping are excellent indicators of weed suppression by the cover crop (fig. 6.1). The choice of cover crop variety, seeding rate, planting date, and other management factors influence the growth of cover crops and their ability to suppress weed growth and weed seed production.

This chapter focuses primarily on how growers can minimize and control weeds during cover cropping periods in the tillage-intensive vegetable systems most common in California. Careful attention to weed management during the cover cropping period is critical because many important weed species occur year-round and vegetable production can occur throughout the year in many regions. The potential biofumigation effects of soil-incorporated cover crop residue in subsequent vegetables, the use of cover crop mulch in reduced-tillage systems, and cover crops as weeds are also briefly addressed.

Weed Control in the Cover Crop

Weed control in cover crops is influenced by several factors (table 6.1). Many of these factors (also discussed in chapter 10) impact other important benefits of cover crops, such as total biomass production. Many weeds that grow during cover cropping mature much faster than the cover crop and can produce viable seed before cover crop termination. The critical period for weed suppression by the cover crop is typically during the first 30 days of cover crop growth. A cover crop's ability to suppress weeds is generally correlated with the cover crop's early-season biomass production rather than with cover crop biomass at maturity. Cover crops such as mustards and cereal rye that maximize light interception with a dense canopy early in the season are often the best at suppressing weed growth and weed seed production.

Table 6.1. Effect of selected factors on weed suppression during cover crop production

Factor	Effect on weed suppression in the cover crop
cover crop variety or mixture	Varieties and mixtures that rapidly develop a canopy are more weed suppressive.
seeding rate	Up to a certain point, higher seeding rates are more weed suppressive than lower rates.
planting date	Earlier fall planting dates allow winter cover crops to germinate quickly and rapidly cover the soil and improves weed suppression.
row spacing	Narrow spacing (6 inches between rows) minimizes competition between cover crop plants and maximizes weed suppression.
irrigation	Irrigation hastens cover crop germination and early canopy development and can increase weed suppression.
planting method	Drilling versus broadcasting a cover crop results in more uniform planting depth, even plant spacing, and an even stand that is more weed suppressive.

Figure 6.1. Burning nettle *(Urtica urens)* weeds that grew under cover crops with increasing weed suppressive abilities: from left to right, legume-oat mixture, oat, cereal rye, and mustard. These plants were harvested 84 days after cover crop planting. Emergence of weeds in the cover crops was not affected by cover crops.
E. Brennan

Figure 6.2. Burning nettle growth in a mixture of 90% legumes and 10% rye planted at 375 pounds per acre (A) versus 125 pounds per acre (B). These photographs were taken at 55 days after planting in a long-term organic systems trial in Salinas, California, that has shown that weed densities in vegetable crops are several times lower where the winter cover crop was grown at the higher seeding rate for several seasons. E. Brennan

The factors in table 6.1 affect each other and can be manipulated to improve the competitive ability of a given cover crop variety or mixture. This is illustrated in a cover crop mixture of 35% bell bean, 25% pea, 15% common vetch, 15% purple vetch, and 10% oat (percentages by seed weight). Mixtures like this are common on organic vegetable farms in the Central Coast region. At a typical seeding rate of 100 pounds per acre, this mixture results in a density of 14 plants per square foot. This mix has poor weed suppressive ability at this seeding rate because the plant density is relatively low. Doubling the seeding rate increases weed suppression, but the increased weed suppression would be more likely to occur if the cover crop were planted with a drill using narrow (6-inch) versus wide (12-inch) spacing between the rows. Narrow spacing increases cover crop competition with weeds earlier in the season because the weeds and cover crops are closer to each other. Adding more oat to the mixture also improves its weed suppression, but it is also likely to suppress the growth of the legume. More research is needed to determine the optimal seeding rates and plant densities that maximize weed suppression by sole and mixed cover crops. With sole cover crops (e.g., rye, mustard, oat, etc.), seeding rates necessary to give good weed control will likely be at least 25 to 50% higher than rates recommended when these crops are grown for seed production. Recent research on the Central Coast found that seeding rates of at least 200 pounds per acre may be necessary to achieve weed suppression by common mixes such as the 90% legume and 10% cereal mix described above (Brennan, unpublished data). A long-term organic trial in Salinas has recorded excellent weed suppression with a cover crop mixture of 90% legume and 10% rye when it was planted at 375 pounds per acre compared with a standard seeding rate of 125 pounds per acre (fig. 6.2); furthermore, weed densities in subsequent vegetables were several times lower where the higher cover crop seeding rate was used (Brennan, unpublished data). Areas of a field that are often more weedy, such as the field edges, should be planted at higher seeding rates by making double passes with the planting equipment.

Cultivation is seldom used to control weeds in cover crops, but it may be an effective and practical option when competition alone is not adequate, and when weather conditions permit cultivation. Row crop cultivators can be used to cultivate between rows of cover crops that are planted on beds, and blind tillage (shallow cultivation over an entire field without regard to row position) implements can be used in cover crops planted in solid stands. This technique selectively uproots weed seedlings at the white thread stage that are below the soil surface and are smaller and more shallowly rooted than the larger-seeded cover crops. Rotary hoes (fig. 6.3) or flex-tine weeders are examples of implements used for blind tillage. With blind tillage, relatively fast (8 to 10 mph) multiple passes can be made to uproot and kill weeds with surprisingly little damage to the young cover crop. Weed seed production by some species in a mixed cover crop can be reduced by up to 80% with the rotary hoe (fig. 6.4).

Figure 6.3. Rotary hoe implement used for blind cultivation.
E. Brennan

Weed Control after Incorporation of Cover Crops

When brassica cover crops are incorporated into the soil, biologically active compounds are released from the residue that can affect weed seeds in the soil. There has been interest in this biofumigation potential of brassica cover crops on weeds (Brown and Morra 1995). Brassica residues contain glucosinolates, which are converted to volatile isothiocyanates that can be toxic to some plant

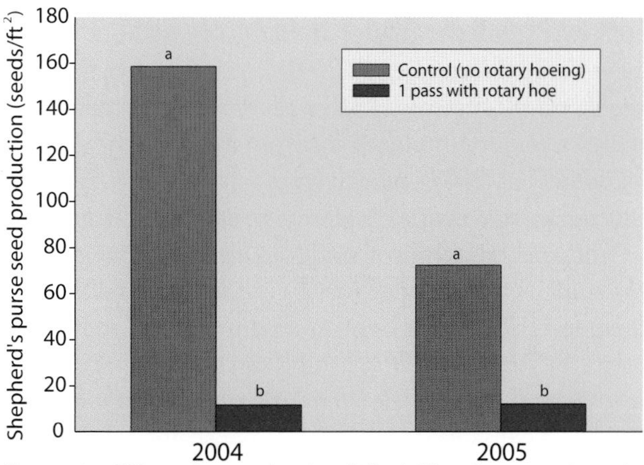

Figure 6.4. Effect of rotary hoeing (blind tillage) on weed seed production by shepherd's purse *(Capsella bursa-pastoris)* in a mixed cover crop (10% cereal rye, 90% legume) grown during the winter in 2 years. Bars topped by different letters within each year are significantly different at an error rate of $P < 0.05$.
Source: Boyd and Brennan 2006

species. The effects of cover crop breakdown products on weeds in field situations are variable, and the mechanisms are poorly understood. For example, in a 2-year study in California, mustard cover crop residue incorporated into the soil reduced weed emergence more than did residue of other cover crops in one year but not in another (Brennan and Smith 2005). It is clear that incorporating residue from a wide range of cover crops can delay seedling emergence and establishment of some weed and crop species: Haramoto and Gallandt (2005) found that the effect of brassica cover crop residue on weed seeds was not different from the effect of residue from a wide range of cover crops.

In addition to the effect of species-specific allelopathic compounds on weeds, adding carbon-rich cover crop biomass to the soil can increase microbial (bacterial and fungal) activity that may influence weed seedling survival and degradation of the weed seed bank, although this is poorly understood (see Kremer 1993; Davis 2007). Soil-incorporated cover crop residue can also alter the amount of available nutrients such as nitrogen that can influence the growth of weeds. For example, highly lignified residues from cereal cover crops can immobilize nitrogen, which may reduce weed growth more than residues from nitrogen-rich legume cover crops that may improve weed growth.

Weed Control by Cover Crop Mulches

Cover crop mulches are seldom used in vegetable production in California, and most of the research on mulches for weed control is from other regions of the United States. In a study of cover crop mulch with processing tomatoes in California, early-season weed control by cover crop mixes of rye-vetch and triticale-vetch was similar to that of an herbicide without a cover crop (Herrero et al. 2001). Weed growth in organic no-till systems is often several times higher than where cover crop mulch is incorporated into the soil (Madden et al. 2004). Mulch from a cowpea cover crop reduced weed densities in desert pepper production in California (Hutchinson and McGiffen 2000). Cover crop mulch can influence weed germination and emergence by altering light, soil temperature, and soil moisture (Teasdale 1996). Mulch can also suppress weeds by releasing allelochemicals that inhibit weed establishment, and if the mulch is dense enough it may physically obstruct weed seedling emergence. Weed control by cover crop mulch depends on weed species and the type and amount of mulch; control generally increases with increasing amounts of mulch. Several times more mulch than would naturally be produced on a given area of land may be required to reduce weed emergence by more than 90% for many weeds. The use of cover crop mulch or surface residue to control weeds may work with direct-seeded vegetables with large seeds, but it is most practical where vegetables are transplanted by hand or with mechanical transplanters that can operate in high-residue situations.

Cover Crops as Weeds

Cover crops can become weeds in subsequent vegetable crops if they mature and set viable seed prior to termination. This is more likely with winter cover crops where wet spring conditions can postpone the cover crop termination and incorporation dates. Growers can minimize the risk of cover crops becoming weeds by monitoring flowering and seed development, selecting cover crop varieties that mature late, avoiding species with hard seed that may persist for many years, and adjusting the planting date to prolong the vegetative growth phase. Mowing the cover crop may prevent or minimize cover crop seed production even when soil conditions are too wet to incorporate the residue.

References

Boyd, N. S., and E. B. Brennan. 2006. Weed management in a legume-cereal cover crop with the rotary hoe. Weed Technology 20:733–737.

Brennan, E. B., and R. F. Smith. 2005. Winter cover crop growth and weed suppression on the central coast of California. Weed Technology 19:1017–1024.

Brennan, E. B., N. S. Boyd, R. F. Smith, and P. Foster. 2009. Seeding rate and planting arrangement effects on growth and weed suppression of a legume-oat cover crop for organic vegetable systems. Agronomy Journal 101:979–988.

Brown, P. D., and M. J. Morra. 1995. Glucosinolate-containing plant tissues as bioherbicides. Journal of Agricultural and Food Chemistry 43:3070–3074.

Davis, A. S. 2007. Nitrogen fertilizer and crop residue effects on seed mortality and germination of eight annual weed species. Weed Science 55:123–128.

Haramoto, E. R., and E. R. Gallandt. 2005. Brassica cover cropping: I. Effects on weed and crop establishment. Weed Science 53:695–701.

Herrero, E. V., J. P. Mitchell, W. T. Lanini, S. R. Temple, et al. 2001. Use of cover crop mulches in a no-till furrow-irrigated processing tomato production system. HortTechnology 11:43–48.

Hutchinson, C. M., and M. E. McGiffen. 2000. Cowpea cover crop mulch for weed control in desert pepper production. HortScience 35:196–198.

Kremer, R. J. 1993. Management of weed seed banks with microorganisms. Ecological Applications 3:42–52.

Madden, N. M., J. P. Mitchell, W. T. Lanini, M. D. Cahn, E. V. Herrero, et al. 2004. Evaluation of conservation tillage and cover crop systems for organic processing tomato production. HortTechnology 14:243–250.

Teasdale, J. R. 1996. Contribution of cover crops to weed management in sustainable agricultural systems. Journal of Production Agriculture 9:475–479.

7 Steven T. Koike and Krishna V. Subbarao

Soilborne Pathogens

Crop rotation is a critically important consideration in managing soilborne pathogens of vegetables. The planting of particular cash crops and cover crops can influence populations of soilborne pathogens and decrease or increase the risk of disease. Because many soilborne pathogens are long-term, chronic threats to vegetable production (Koike et al. 2003), growers should carefully plan which crops and cover crops to plant.

Benefits of Cover Crops for Managing Pathogens

Cover crops are primarily planted to help manage the soil and improve growing conditions for subsequent agronomic crops. From the disease management perspective, planting a cover crop species that is a nonhost of soilborne pathogens would be a benefit. The selection of an appropriate cover crop would therefore provide the agronomic benefits for conditioning the soil while not allowing soilborne pathogens to increase inoculum and cause problems for the crop to follow. For example, growing grass cover crops such as rye or barley in a lettuce rotation improves soil conditioning and nutrient management and does not increase sclerotia of the lettuce pathogen *Sclerotinia minor*. However, growing vetch cover crops can significantly increase sclerotia of this pathogen.

While growing cover crops is highly recommended, the actual reduction of soilborne pathogens due to direct activity of the cover crop is generally not well documented. A cause-and-effect relationship between cover crops and disease reduction in the cash crop has been clearly demonstrated in only a few cases. What is generally accepted, however, is that cover crops contribute to an improvement in soil microbial diversity, soil conditions for the subsequent crops, and overall soil health. Such benefits should create conditions that favor the crop and perhaps affect soilborne pathogens. Specifically, increased diversity in soil microorganisms can put pathogens at a disadvantage due to greater competition, antagonism, predation, and parasitism.

Demonstrated suppression of soilborne pathogens has been seen with some brassica species. When crop residues from these species decompose, glucosinolate chemicals in the plant tissues break down into substances that are toxic to soilborne pathogens. These plant substances may also trigger changes in the overall soil microbe community and result in further suppression of pathogens. Extensive research with some *Brassica oleracea* crops (broccoli and cabbage) demonstrated that these plants can significantly reduce inoculum of pathogens such as *Verticillium dahliae* and *Sclerotinia minor* (Hao et al. 2003; Koike and Subbarao 2000; Koike et al. 2000; Subbarao et al. 1999). These findings have encouraged some growers to rotate broccoli with susceptible crops (e.g., lettuce) or to plant broccoli as a cover crop to address soilborne problems. However, mustard cover crops such as Indian *(Brassica juncea)* and white *(Sinapis alba)* mustard have not demonstrated the same ability to reduce soilborne diseases for subsequent vegetable crops (Bensen et al. 2009; Hartz et al. 2005). There is some evidence that sudangrass may suppress Verticillium wilt of potato (Davis et al. 1996).

It is likely that a number of important factors influence whether soilborne pathogens can be significantly reduced by cover crops. Of course, a key factor is the choice of cover crop species. In addition, other factors include region where the crops are being grown (coastal or inland, northern or southern), soil type, the microbial communities of the particular soil, and cropping patterns.

Cover Crops as Hosts of Pathogens

In contrast to possible disease management benefits, cover crops can at times increase the inoculum of certain soilborne pathogens. Cover crop selection must therefore take into account possible adverse effects on subsequent vegetable plantings. Phacelia, Austrian pea, and vetch are hosts to *Sclerotinia minor* and could increase lettuce drop on subsequent lettuce crops (Dillard and Grogan 1985; Koike et al. 1996) (figs. 7.1, 7.2). Vetches can also cause *Pythium* and *Rhizoctonia* populations to increase (Sumner et al. 1995), which can create damping-off diseases in

Figure 7.1. Sclerotia of *Sclerotinia minor* forming inside the stem of an infected phacelia plant. Some cover crop plantings can cause significant increases in pathogen inoculum. S. Koike

some vegetable crops. Velvet bean cover crop residues left on the soil surface have been shown to increase wirestem problems (caused by *Rhizoctonia solani*) for collards (Keinath et al. 2003). Growing any brassica cover crop may simultaneously increase clubroot risk (caused by *Plasmodiophora brassicae*) for subsequent crucifer crops while also suppressing other soilborne pathogens (Smith et al. 2005) (fig. 7.3). These examples demonstrate that a particular cover crop can be beneficial for one vegetable crop species but create significant problems for a different vegetable crop. The same considerations hold for foliar diseases that affect cover crops and might spread to nearby agronomic crops; the recent finding that *Impatiens necrotic spot virus* can infect both fava bean cover crops and nearby lettuce and radicchio plantings illustrates such concerns (fig. 7.4). Mustard cover crops are susceptible to the same downy mildew *(Peronospora parasitica)* that infects vegetable brassica crops (fig. 7.5).

Cover crop effects in increasing pest populations can be even more complex when considering multiple pests in vegetable production systems. The cover crop selection may not influence soil fungi but could increase other soil pests such as nematodes. This complexity means that the decision-making process of selecting a cover crop may not be a simple one. Oilseed radish, for example, offers the advantages of not being a host for *Sclerotina minor* and being a potential trap crop for the sugarbeet cyst nematode *(Heterodera schachtii)*. However, this cover crop species is a host of the clubroot pathogen and two species of root-knot nematodes *(Meloidogyne incognita* and *M. javanica)* (Gardner and Caswell-Chen 1994). Phacelia is not a host to cyst nematode, but it is susceptible to root-knot

Figure 7.2. Vetch stems infected with *Sclerotinia minor*. Diseased vetch plantings can significantly increase *S. minor* populations and cause problems for subsequent lettuce plantings. R. Smith

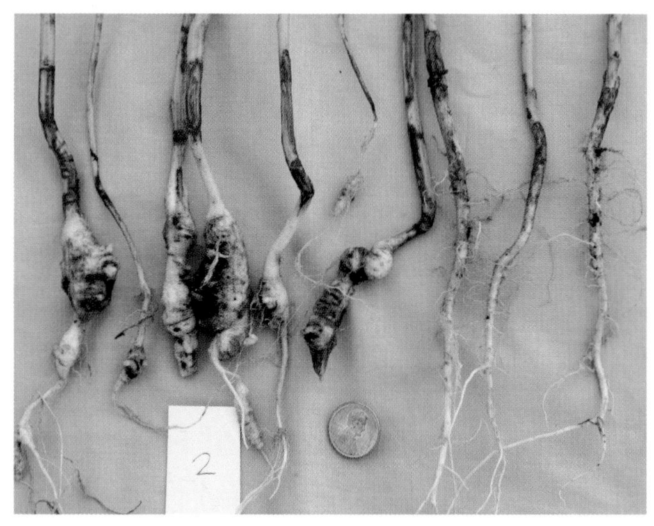

Figure 7.3 Mustard cover crop plants infected with *Plasmodiophora brassicae*. Healthy plants are on the right. This soilborne pathogen can increase on cruciferous cover crops and cause increased problems for broccoli, cauliflower, and other crucifers. S. Koike

Figure 7.4 Fava bean infected with *Impatiens necrotic spot virus*. The thrips vector can transmit this virus from the cover crop to adjacent hosts such as lettuce and radicchio. S. Koike

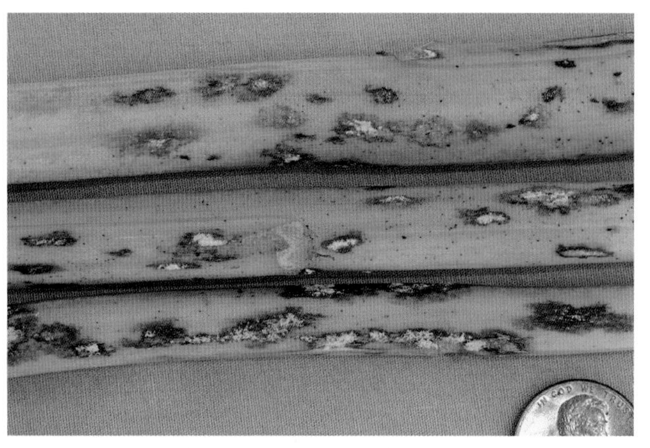

Figure 7.5 Mustard cover crop stems infected with the downy mildew pathogen *Peronospora parasitica*. The spores forming on the cover crop can spread to adjacent crucifer plantings. S. Koike

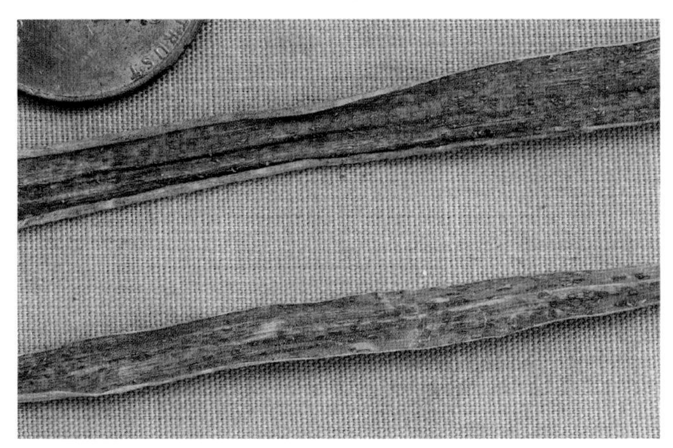

Figure 7.6. Rye cover crop leaves infected with *Puccinia graminis* f. sp. *secalis*. This rust pathogen is host-specific to rye and does not infect agronomic crops. S. Koike

Figure 7.7. Vetch cover crop leaves infected with *Ramularia sphaeroidea*. This pathogen is host-specific to vetch and does not spread to agronomic crops. S. Koike

nematode (Gardner and Caswell-Chen 1993) and is a host, as mentioned above, of the lettuce drop pathogen *Sclerotina minor*. Vetch species are also hosts of *S. minor* and *Meloidogyne* spp. (Johnson et al. 1992). For more information on cover crop and nematode interactions, see chapter 8.

A final consideration is the direct effect plant pathogens have on the cover crops. In some cases, the pathogen infects only the cover crop and has an impact on cover crop performance and growth but does not infect the vegetable crops that follow. For example, if environmental conditions are favorable for disease development, the foliar rust disease caused by *Puccinia graminis* f. sp. *secalis* can cause severe leaf necrosis and growth reduction of Merced rye plantings (Smith et al. 2003) (fig. 7.6). This rust, however, is restricted to rye and will not affect vegetable crops. Similarly, *Ramularia sphaeroidea* causes a leaf spot on vetch but will not cause any disease on lettuce or other crops to follow (Koike et al. 2004) (fig. 7.7). Interestingly, if Ramularia leaf spot is severe on the vetch, the vetch can defoliate and cover the ground with dead leaves. We have observed that sclerotia of *Sclerotina minor* can colonize these senescent tissues and reproduce; therefore, Ramularia leaf spot may have an indirect effect in increasing lettuce drop problems on lettuce.

Summary

- When choosing cover crop species, evaluate the degree of cover crop susceptibility to pathogens and other pests and the relative importance of these problems for subsequent vegetable crops (table 7.1). If the primary concern is lettuce, cover crop effects on *Pythium* populations will be less important than effects on *Sclerotinia*. Therefore, the cash crop rotation will help dictate which cover crop species should be used.

Table 7.1. Effect of selected cover crops on soilborne pathogens and nematodes

Cover crop	Positive effect	Negative effect
Austrian pea	—	host of *Sclerotinia minor*
merced rye	nonhost of vegetable pathogens	—
mustard	suppress pathogens via glucosinolates	host of *Sclerotinia minor;* host of *Plasmodiophora*
oilseed radish	trap crop for *Heterodera* nematode	host of *Plasmodiophora;* host of *Meloidogyne* nematode
phacelia	nonhost of *Heterodera* nematode	host of *Sclerotinia minor;* host of *Meloidogyne* nematode
vetch	—	host of *Sclerotinia minor*; increases *Pythium, Rhizoctonia;* host of *Meloidogyne* nematode

- Keep accurate records of soilborne pathogens known to be a problem in fields. A documented history of the pathogens present will help guide the selection of cover crops in such locations.
- Inspect the cover crops as they grow. Because cover crops may play a role in the epidemiology of soilborne diseases, growers should be aware of disease symptoms developing during cover crop growth. This would entail periodic monitoring of the cover crop. Should disease symptoms develop on cover crops, send representative samples to extension personnel, diagnostic labs, and other professionals who can diagnose the disease.
- Keep note also of foliar problems that occur on the cover crops. In some situations, such developments might also impact cash crops.

References

Bensen, T. A., R. F. Smith, K. V. Subbarao, S. T. Koike, S. A. Fennimore, and S. Shem-Tov. 2009. Mustard and other cover crop effects vary on lettuce drop caused by *Sclerotinia minor* and on weeds. Plant Disease 93:1019–1027.

Davis, J. R., O. C. Huisman, D. T. Westermann, S. L. Hafez, et al. 1996. Effects of green manures on Verticillium wilt of potato. Phytopathology 86:444–453.

Dillard, H. R., and R. G. Grogan. 1985. Influence of green manure crops and lettuce on sclerotial populations of *Sclerotinia minor*. Plant Disease 69:579–582.

Gardner, J., and E. P. Caswell-Chen. 1993. Penetration, development, and reproduction of *Heterodera schachtii* on *Fagopyrum esculentum, Phacelia tanacetifolia, Raphanus sativus, Sinapis alba,* and *Brassica oleracea.* Journal of Nematology 25:695–702.

———. 1994. *Raphanus sativus, Sinapis alba,* and *Fagopyrum esculentum* as hosts to *Meloidogyne incognita, Meloidogyne javanica,* and *Plasmodiophora brassicae*. Journal of Nematology 26:756–760.

Hao, J. J., K. V. Subbarao, and S. T. Koike. 2003. Effects of broccoli rotation on lettuce drop caused by *Sclerotinia minor* and on the population density of sclerotia in soil. Plant Disease 87:159–166.

Hartz, T. K., P. R. Johnstone, E. M. Miyao, and R. M. Davis. 2005. Mustard cover crops are ineffective in suppressing soilborne disease or improving processing tomato yield. HortScience 40:2016–2019.

Johnson, A. W., A. M. Golden, D. L. Auld, and D. R. Sumner. 1992. Effects of rapeseed and vetch as green manure crops and fallow on nematodes and soilborne pathogens. Journal of Nematology 24:117–126.

Keinath, A. P., H. F. Harrison, P. C. Marino, D. M. Jackson, and T. C. Pullaro. 2003. Increase in populations of *Rhizoctonia solani* and wirestem of collard with velvet bean cover crop mulch. Plant Disease 87:719–725.

Koike, S. T., and K. V. Subbarao. 2000. Broccoli residues can control Verticillium wilt of cauliflower. California Agriculture 54(3): 30–33.

Koike, S. T., R. F. Smith, L. E. Jackson, L. J. Wyland, et al. 1996. Phacelia, lana woollypod vetch, and Austrian winter pea: Three new cover crop hosts of *Sclerotinia minor* in California. Plant Disease 80:1409–1412.

Koike, S. T., C.-L. Xiao, J. C. Hubbard, K. F. Schulbach, and K. V. Subbarao. 2000. Effects of broccoli residue on the cauliflower–*Verticillium dahliae* host-pathosystem. In E. C. Tjamos et al., eds., Advances in Verticillium: Research and disease management. St. Paul: American Phytopathological Society Press. 317–321.

Koike, S. T., K. V. Subbarao, R. M. Davis, and T. A. Turini. 2003. Vegetable diseases caused by soilborne pathogens. Oakland; University of California Agriculture and Natural Resources Publication 8099. ANR Communication Services Web site, http://anrcatalog.ucdavis.edu/DiseasesDisorders/8099.aspx.

Koike, S. T., R. F. Smith, P. W. Crous, and J. Z. Groenewald. 2004. Leaf and stem spot, caused by *Ramularia sphaeroidea*, on purple and Lana woollypod vetch (*Vicia* spp.) cover crops in California. Plant Disease 88:221.

Smith, R., E. Brennan, S. T. Koike, and A. Barber. 2003. Rust on Merced rye cover crops evaluated. Monterey County Crop Notes. September.

Smith, R., S. T. Koike, G. Colfer, T. Bensen, and K. Kammeijer. 2005. Clubroot evaluation on mustard cover crops. Monterey County Crop Notes. March–April.

Subbarao, K. V., J. C. Hubbard, and S. T. Koike. 1999. Evaluation of broccoli residue incorporation into field soil for Verticillium wilt control in cauliflower. Plant Disease 83:124–129.

Sumner, D. R., S. C. Phatak, J. D. Gay, R. B. Chalfant, K. E. Brunson, and R. L. Bugg. 1995. Soilborne pathogens in a vegetable double crop with conservation tillage following winter cover crops. Crop Protection 14:495–500.

8 *Howard Ferris*

Plant and Soil Nematodes

Soil nematodes are microscopic worms that range in length from about 0.2 to 5.0 millimeters, with the majority less than 1 millimeter (fig. 8.1). Nematodes are among the simplest of multicellular organisms. They are very abundant in soil and aquatic systems and by some estimates are the most abundant multicellular animals on the planet; they exhibit a great diversity, in both form and function. Certain plant-feeding nematodes are important pests in agriculture (fig. 8.2), but many of the nematode species in soil feed on bacteria, fungi, algae, other nematodes, or other small invertebrates and contribute to the ecosystem services provided by the soil food web. The food web is the community of soil organisms that are interdependent for sources of food. Among the services provided by the web are nutrient cycling, decomposition of organic matter, and regulation of soil pest species.

Many species of plant-feeding nematodes are associated with vegetable crops in California (table 8.1). They vary in their feeding habits, the nature and amount of damage they cause to the plant root system, their rates of population increase on a favorable host, and their rates of population decline in the absence of a host. All plant-feeding nematodes insert stylets into plant cells to withdraw cell contents. Sedentary endoparasites (fig. 8.3) enter a root, establish a feeding site, and do not move from that site; they often cause galls and deposit eggs in egg masses on the outside of roots. Sedentary semiendoparasites partially enter a root and establish a feeding site, but the majority of the body remains on the root surface; they do not move from the feeding site and may retain eggs in the body or deposit them in egg masses. Migratory endoparasites enter a root, feed from individual

Figure 8.2. Severe galling of a tomato root in response to infection by root-knot nematodes, *Meloidogyne* spp. Details of a gall and nematode feeding site are provided in figure 8.3. J. K. CLARK

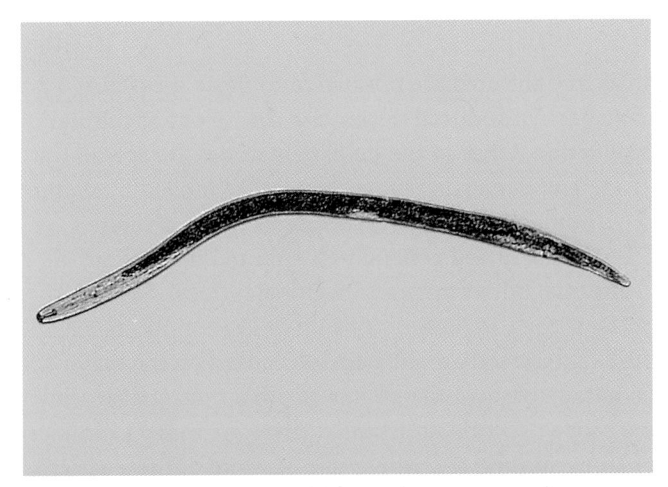

Figure 8.1. An adult female of the root lesion nematode *Pratylenchus penetrans*. This nematode is a migratory endoparasite about 0.5 mm long: it enters the root cortex and moves from cell to cell. The head of the nematode is at the bottom left of the photograph. The small object resembling a pin at the head end is the stylet, which is used for penetrating cell walls and extracting cell contents. SOCIETY OF NEMATOLOGISTS

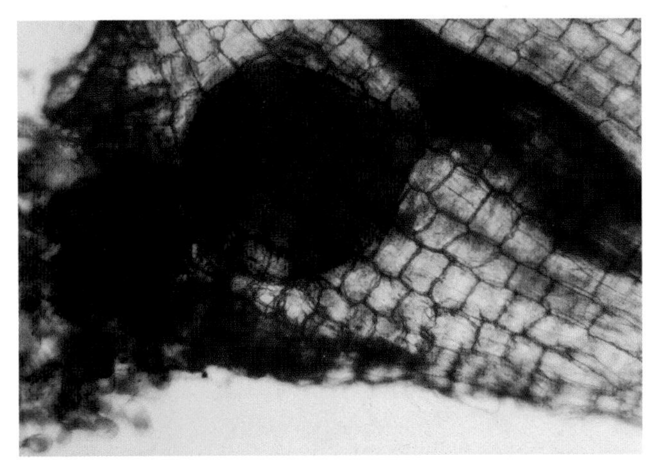

Figure 8.3. Section of a root with central vascular tissues distorted and disrupted by giant cell feeding sites of a root-knot nematode. The root cortex is also swollen, resulting in the galled appearance of the root system shown in fig. 8.2. In this photograph, the body of the swollen female nematode, and her eggs in a gelatinous mass on the outside of the root, have been stained red for contrast. R. S. HUSSEY

Table 8.1. Nematode pests of vegetable crops in California

Common name	Scientific name	Distribution	Feeding habits	Vegetable hosts	Examples of resistant or nonhost cover crops*
NEMATODES OF MAJOR IMPORTANCE					
root-knot nematodes	*Meloidogyne arenaria*	statewide	sedentary endoparasites	many	many grasses and grains, some legumes
	Meloidogyne chitwoodi				many grasses, grains, and brassicas
	Meloidogyne hapla				some grasses and legumes
	Meloidogyne incognita				some legumes, composites, and brassicas
	Meloidogyne javanica				some grasses, legumes, composites, and brassicas
cyst nematodes	*Heterodera cruciferae*	Salinas Valley	sedentary semiendoparasites	cole crops	
	Heterodera schachtii	statewide			some legumes and a few brassicas
lesion nematodes	*Pratylenchus brachyurus*	statewide	migratory endoparasites	many	some legumes and sorghum
	Pratylenchus coffeae				a few compositae
	Pratylenchus crenatus				some alfalfa cultivars
	Pratylenchus neglectus				some alfalfa and clover cultivars
	Pratylenchus penetrans				a few grasses
	Pratylenchus scribneri				rape
	Pratylenchus thornei				turnip and red clover
	Pratylenchus zeae				sunflower and cowpea
needle nematode	*Longidorus africanus*	desert valleys	ectoparasite, root tip feeder	many	some brassicas
NEMATODES USUALLY OF LESSER IMPORTANCE					
dagger nematode	*Xiphinema americanum*	statewide	ectoparasite, root tip feeder	many	—
pin nematode	*Paratylenchus* spp.	statewide	ectoparasites	many	—
ring nematode	*Criconemoides* spp.	statewide	ectoparasites	many	—
spiral nematodes	*Helicotylenchus* spp.	statewide	ectoparasites and migratory endoparasites	many	—
	Rotylenchus robustus	Central Coast	ectoparasite		
sheath nematodes	*Hemicycliophora* spp. *Hemicriconemoides* spp.	patchy in wet, sandy soils	ectoparasites, root tip feeders	many	—
stubby-root nematodes	*Trichodorus* spp. *Paratrichodorus* spp.	statewide	ectoparasites, root tip feeders	many	—
stunt nematodes	*Tylenchorhynchus* spp. *Merlinius* spp.	statewide	ectoparasites	many	—

Note:
*See Nemabase or the cover crop selector software in Nemaplex for species or cultivars that have the desired characteristics of resistance or nonhost status. Nematodes that have only a few or no examples listed in this table either have a wide host range or few cover crops have been tested against them. In these cases, employ Strategy 2 (see text).

cells, and continue to move through and even out of the root tissue; they deposit eggs individually either in root tissues or in soil. Ectoparasites (fig. 8.4) remain on the outside of the root and, depending on the length of the stylet, may feed from cells on the root surface or deep in root tissues. For the purpose of this chapter, a host is a plant on which a nematode species is able to feed and reproduce. In that case, if soil temperature conditions are favorable for nematode reproduction, there will be more nematodes at the end of the crop season than at the beginning.

Nematodes are easily moved from field to field on equipment and vehicles, in soil associated with planting material, inside vegetative planting material such as bulbs or tubers, and in the roots of contaminated transplants or nursery stock. In fact, many of the important nematode species in vegetable production areas in California are not native to those areas but have been introduced during the course of agricultural production. Once in the field, nematodes are spread by tillage and harvesting practices and are moved by water.

Some of the nematode species listed in table 8.1 are widely and generally distributed throughout California; others are more regional or local. Several of the species indicated as being of lesser importance probably exacerbate the damage caused by the major nematode pests or the effects of other crop stresses. In the scope of world agriculture, there are many examples of crops and regions where one or more of these latter species are of major importance, and this could also be the case under local conditions in California. Further, it is highly likely that when effective nematicides are less readily available, we will see a greater diversity of nematode problems in vegetable crops.

Selecting Cover Crop Species for Nematode Management

Cover crops differentially affect many components of the soil food web. There are several mechanisms through which they may reduce populations of plant-feeding nematodes or reduce their potential damage to subsequent vegetable crops. Some of the mechanisms are more clearly defined and understood than others, and most cover-cropping approaches probably encompass more than one of the mechanisms. As a caveat, most plant species are hosts to several plant-feeding nematode species. A cover crop designed to reduce one nematode species may increase another, so it is important to monitor the response of the nematode community.

It is quite difficult to find cover crops that provide agronomic benefits, grow at a convenient time of the year in relation to the crop season, and are nonhosts to important nematode species. For example, velvet bean *(Mucuna deeringiana)* is a tropical legume that is a nonhost to the root-knot nematode *Meloidogyne incognita*, but it would not function as a winter cover crop in California. Conversely, many of the plant species commonly suggested as cool-season cover crops in California are actually weak to moderate hosts of root-knot nematodes. Consequently, two strategies should be considered in growing cover crops, whether for nematode management or not (fig. 8.5). The strategies are based on seasonal differences in soil temperature and the fact that nematodes, as cold-blooded organisms, have metabolic rates that respond to changes in temperature. Most California nematode populations for which thermal requirements have been determined cease to move at soil temperatures below 55°F and become metabolically inactive below 50°F.

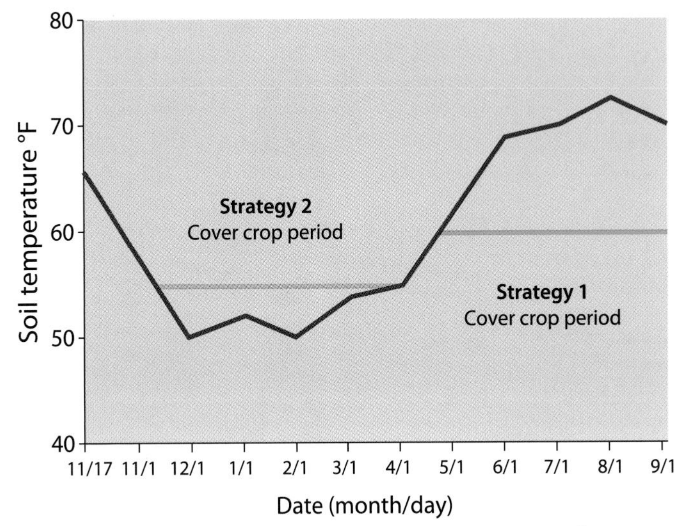

Figure 8.5. Example of seasonal soil temperature ranges for two strategies of using cover crops to reduce population levels of plant-feeding nematodes.

Strategy 1: Grow a nonhost cover crop or cover crop mixture when soil temperatures are at levels at which nematodes will be active (60°F and above). Population levels of plant-feeding nematodes decline because they are metabolically active but unable to replenish energy reserves by feeding on the nonhost.

Strategy 2: Grow cool-season cover crops that may be weak to moderate hosts of the target nematode species; sow them when soil temperatures drop below 55°F (when nematode activity is very low) and terminate them when soil temperatures increase above that threshold in the spring. If the cover crop will span both soil temperature regimes, for example, from February to June (see fig. 8.5), select varieties that are nonhosts to the target nematodes.

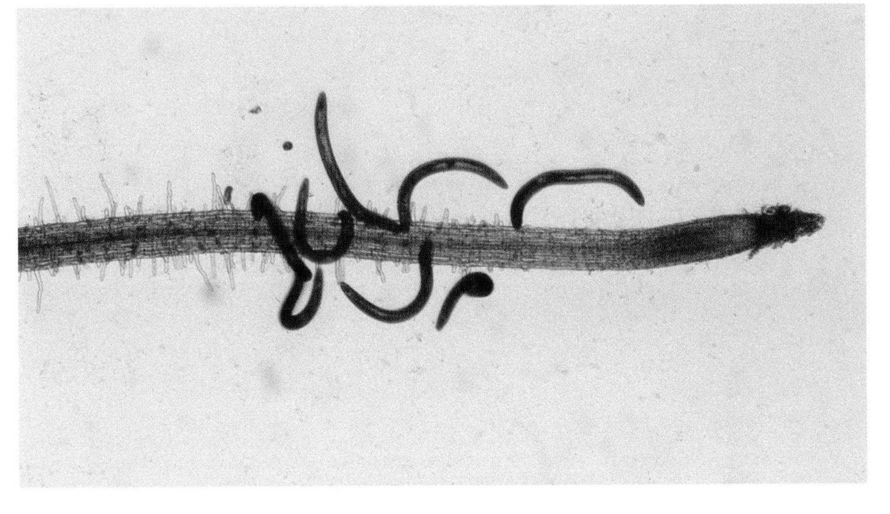
Fig. 8.4. Plant-parasitic nematodes, in this case ectoparasites which do not enter the root system, attracted to feeding sites just behind the root tip. R. S. Hussey

Table 8.2. Host status of cover crops for selected nematode species

Cover crop	Common varieties*	*Meloidogyne arenaria*	*Meloidogyne chitwoodi*	*Meloidogyne hapla*	*Meloidogyne incognita*	*Meloidogyne javanica*	*Heterodera cruciferae*	*Heterodera schachtii*
		Root-knot nematode	Root-knot nematode	Root-knot nematode	Root-knot nematode	Root-knot nematode	Crucifer cyst nematode	Sugarbeet cyst nematode
GRASSES								
Avena sativa oat	Cayuse Swan, Canota	—	■	■	■	■	—	●
Bromus mollis brome	Blando	—	—	■	●	—	—	—
Hordeum vulgare barley	UC 603, UC 937	■	■	●	■	■	●	●
Secale cereale cereal rye	Merced	■	■	—	■	■	—	—
Sorghum bicolor ssp. *drummondii* sudangrass	Piper	●	●	—	●	●	—	—
Sorghum bicolor × *S. bicolor* ssp. *drummondii* sorghum-sudangrass	Sordan 79	●	●	—	●	●	—	—
Triticosecale rimpaui triticale	Juan	■	—	—	■	—	—	—
Triticum aestivum wheat	Summit, Dirkwin	■	■	●	■	■	■	●
LEGUMES								
Crotalaria spectabilis sunn hemp	Tropic Sun	●	—	—	●	●	—	—
Medicago sativa alfalfa		▲	●	■	●	■	—	●
Pisum sativum field pea	Magnus, Biomaster, Dundale	■	■	■	■	■	—	●
Trifolium spp. clover		■	■	■	■	■	—	●
Trifolium pratense red clover	various	■	■	■	■	■	—	●
Vicia benghalensis purple vetch	various	▲	—	—	—	—	—	■
Vicia faba bell bean	various	■	—	■	■	■	—	—
Vicia sativa common vetch	various	●	—	—	■	■	—	—
Vicia villosa hairy vetch	various	■	—	■	▲	—	—	—
Vicia villosa ssp. *dasycarpa* woollypod vetch	Lana	—	—	—	—	—	—	—
Vigna unguiculata cowpea	CB5, Iron Clay, Chinese Red	●	●	■	●	■	—	●
BRASSICACEAE								
Brassica juncea Indian mustard	Pacific Gold, Caliente 99	■	—	■	■	—	■	■
Brassica napus canola	Humus, Erica	—	—	■	■	■	■	■
Raphanus sativus oilseed radish	Nematrap	■	●	■	■	■	■	■
Sinapis alba white mustard	Ida Gold	■	—	—	■	■	■	■
OTHERS								
Fagopyrum esculentum buckwheat	various	■	—	●	■	■	—	●
Phacelia tanacetifolia phacelia	Phaci	—	—	—	■	■	—	—

KEY: ■ = host ● = resistant ▲ = moderately resistant — = No information available.

Notes: The information in the table is extracted from the world literature and from local observations. There may be differences among cultivars or local nematode populations with regard to the host status characteristics. Please refer to Nemabase or the cover crop selector software in Nemaplex (see text) for species or cultivars that have the desired characteristics of resistance or nonhost status. In cases where the cover crop is a host and can be grown in the cool season, Strategy 2 (see text) should be employed.

**Check with local seed distributors for varieties appropriate to your region.*

Table 8.2. Host status of cover crops for selected nematode species *(continued)*

Cover crop	Common varieties*	*Pratylenchus brachyurus* Root lesion nematode	*Pratylenchus coffeae* Root lesion nematode	*Pratylenchus crenatus* Root lesion nematode	*Pratylenchus neglectus* Root lesion nematode	*Pratylenchus penetrans* Root lesion nematode	*Pratylenchus scribneri* Root lesion nematode	*Pratylenchus thornei* Root lesion nematode
GRASSES								
Avena sativa oat	Cayuse Swan, Canota	■	—	■	■	■	—	■
Bromus mollis brome	Blando	—	—	—	—	—	—	—
Hordeum vulgare barley	UC 603, UC 937	■	—	■	■	■	■	■
Secale cereale cereal rye	Merced	■	—	■	■	■	—	■
Sorghum bicolor × *S. bicolor* ssp. *drummondii* sorghum-sudangrass	Sordan 79	—	—	—	—	—	■	—
Sorghum bicolor ssp. *drummondii* sudangrass	Piper	■	—	—	■	■	■	■
Triticosecale rimpaui triticale	Juan	—	—	—	—	—	—	■
Triticum aestivum wheat	Summit, Dirkwin	■	—	—	■	■	■	■
LEGUMES								
Crotalaria spectabilis sunn hemp	Tropic Sun	■	—	—	—	■	—	—
Medicago sativa alfalfa	various	▲	■	●	■	■	●	■
Pisum sativum field pea	Magnus, Biomaster, Dundale	—	■	■	■	■	—	■
Trifolium spp. clover	various	■	■	■	■	■	■	■
Trifolium pratense red clover	various	■	■	■	■	■	■	●
Vicia benghalensis purple vetch	various	—	—	—	—	—	—	●
Vicia faba bell bean	various	—	—	—	■	■	—	■
Vicia sativa common vetch	various	—	—	—	—	■	—	■
Vicia villosa hairy vetch	various	▲	—	—	—	■	—	—
Vicia villosa ssp. *dasycarpa* woollypod vetch	Lana	—	—	—	—	—	—	—
Vigna unguiculata cowpea	CB5, Iron Clay, Chinese Red	■	—	—	■	■	—	■
BRASSICACEAE								
Brassica juncea Indian mustard	Pacific Gold, Caliente 99	—	—	—	—	—	—	—
Brassica napus canola	Humus, Erica	—	—	—	■	■	●	—
Raphanus sativus oilseed radish	Nematrap	—	—	—	—	■	■	▲
Sinapis alba white mustard	Ida Gold	—	—	—	—	■	—	—
OTHERS								
Fagopyrum esculentum buckwheat	various	■	—	—	—	■	—	—
Phacelia tanacetifolia phacelia	Phaci	—	—	—	—	—	—	—

KEY: ■ = host ● = resistant ▲ = moderately resistant — = No information available.

Notes: The information in the table is extracted from the world literature and from local observations. There may be differences among cultivars or local nematode populations with regard to the host status characteristics. Please refer to Nemabase or the cover crop selector software in Nemaplex (see text) for species or cultivars that have the desired characteristics of resistance or nonhost status. In cases where the cover crop is a host and can be grown in the cool season, Strategy 2 (see text) should be employed.

**Check with local seed distributors for varieties appropriate to your region.*

Table 8.2. Host status of cover crops for selected nematode species *(continued)*

		Pratylenchus zeae	*Longidorus africanus*	*Xiphinema americanum*	*Paratylenchus* spp.	*Criconemoides* spp.	*Helicotylenchus* spp.	*Rotylenchus robustus*
Cover crop	**Common varieties**[a]	**Root lesion nematode**	**Needle nematode**	**Dagger nematode**	**Pin nematode**	**Ring nematode**	**Spiral nematode**	**Spiral nematode**
GRASSES								
Avena sativa oat	Cayuse Swan, Canota	—	■	■	■	■	■	—
Bromus mollis brome	Blando	—	—	■	—	▲	—	—
Hordeum vulgare barley	UC 603, UC 937	—	■	■	■	■	■	—
Secale cereale cereal rye	Merced	■	—	—	■	■	■	—
Sorghum bicolor × *S. bicolor* ssp. *drummondii* sorghum-sudangrass	Sordan 79	—	—	■	—	■	●	—
Sorghum bicolor ssp. *drummondii* sudangrass	Piper	■	■	■	■	■	■	—
Triticosecale rimpaui triticale	Juan	—	—	—	—	—	■	—
Triticum aestivum wheat	Summit, Dirkwin	—	■	■	■	■	■	—
LEGUMES								
Crotalaria spectabilis sunn hemp	Tropic Sun	■	—	●	—	—	■	—
Medicago sativa alfalfa		■	■	■	■	■	■	—
Pisum sativum field pea	Magnus, Biomaster, Dundale	—	■	—	●	—	■	■
Trifolium spp. clover	various	—	—	■	■	■	■	—
Trifolium pratense red clover	various	—	—	■	■	■	■	—
Vicia benghalensis purple vetch	various	—	—	—	—	—	—	—
Vicia faba bell bean	various	—	—	—	—	—	—	—
Vicia sativa common vetch	various	—	—	—	—	—	—	—
Vicia villosa hairy vetch	various	—	—	—	■	■	—	—
Vicia villosa ssp. *dasycarpa* woollypod vetch	Lana	—	—	—	—	—	—	—
Vigna unguiculata cowpea	CB5, Iron Clay, Chinese Red	●	—	■	●	■	■	—
BRASSICACEAE								
Brassica juncea Indian mustard	Pacific Gold, Caliente 99	—	—	—	—	—	—	—
Brassica napus canola	Humus, Erica	—	—	—	—	■	■	—
Raphanus sativus oilseed radish	Nematrap	—	▲	—	■	—	■	—
Sinapis alba white mustard	Ida Gold	—	—	—	—	—	—	—
OTHERS								
Fagopyrum esculentum buckwheat	various	—	—	—	—	—	—	—
Phacelia tanacetifolia phacelia	Phaci	—	—	—	—	—	—	—

KEY: ■ = host ● = resistant ▲ = moderately resistant — = No information available.

Notes: The information in the table is extracted from the world literature and from local observations. There may be differences among cultivars or local nematode populations with regard to the host status characteristics. Please refer to Nemabase or the cover crop selector software in Nemaplex (see text) for species or cultivars that have the desired characteristics of resistance or nonhost status. In cases where the cover crop is a host and can be grown in the cool season, Strategy 2 (see text) should be employed.

[a]Check with local seed distributors for varieties appropriate to your region.

The mechanisms below by which cover crops affect nematode populations depend on which of the above strategies is employed.

Mechanism 1: The cover crop or cover crop mixture is comprised of nonhosts that do not provide food for important species of plant-feeding nematodes. This mechanism is particularly effective with Strategy 1 under environmental conditions in which nematodes are metabolically active. Selection of a nonhost cover crop can present a challenge because most soils provide habitat to several species of plant-feeding nematodes, and those nematodes vary in their host ranges. A cover crop that is a nonhost to one species may allow feeding and reproduction of another. Consequently, it is important to know the species of nematodes that are problematic in each field and on the crops that will be grown in that field. There are too many combinations of cover crops, crops, and nematodes to list here, but an enormous amount of host status information is assembled in the Nemabase database (Ferris and Caswell-Chen 1997; Ferris, 1999–2011), which can be accessed online in the Nemaplex Web site, http://plpnemweb.ucdavis.edu/nemaplex, or downloaded from the UC Statewide Integrated Pest Management Web site, http://www.ipm.ucdavis.edu. In some cases, nematode-resistant varieties of the cover crops are available. Detailed information on those varieties is available in Nemabase. Some may not be suited to local conditions, or they may not be locally available, so some research will be necessary. Another database in Nemaplex allows growers to select cover crops or crop sequences that include nonhosts or poor hosts to several nematodes in a field.

Mechanism 2: The cover crop has toxic root exudates, or its breakdown products are toxic to nematodes. This mechanism can be effective under either Strategy 1 or Strategy 2. A wide range of plants have metabolites or residues that are toxic to plant-feeding nematodes, including, for example, sudangrass *(Sorghum sudanense)* (Adewusi 1990; Donkin et al. 1995); rapeseed and mustards (*Brassica* spp.) (Zasada and Ferris 2003, 2004); marigold (*Tagetes* spp.) (Daulton and Curtis 1963; Gommers and Bakker 1988); rye *(Secale cereale)* (Zasada et al. 2005); and sunn hemp *(Crotalaria juncea)* (Wang et al. 2001). In many cases, the mode of action and chemistry of the toxic compounds is not well known. The use of rye as a cover crop is an interesting example. When incorporated into soil, hydroxamic acids in the rye residue degrade to substances of varying toxicity to root-knot and other nematodes. However, the amount of each toxic substance formed is affected by soil temperature and moisture and is somewhat unpredictable (Zasada et al. 2005). Further, rye, while growing, is a reasonably good host to several of the nematodes in table 8.1, so rye is a Strategy 2 cover crop in nematode management, depending on the species present. Various *Brassica* species (certain radishes, rapeseed, and mustards) produce glucosinolates that are hydrolyzed in the presence of the enzyme myrosinase to isothiocyanates, which are nematicidal. The glucosinolates differ among plant species and may be concentrated in different plant parts. The amount of plant biomass that must be incorporated into soil for effective nematode control depends on the types of glucosinolates and their concentrations in the plant material, so success with this strategy is also somewhat unpredictable (Zasada and Ferris 2003, 2004). Nonetheless, the use of cover crops with toxic breakdown products is generally effective in reducing population levels of plant-feeding nematodes.

Mechanism 3: The cover crop is used as a trap crop. This mechanism is most effective under a modification of Strategy 2 and with sedentary endoparasites, such as root-knot and cyst nematodes which feed within the plant root (see Timmermans 2005). Plants can be used as trap crops if they stimulate nematode egg hatch or initiation of the nematode life cycle and can be terminated before the nematodes reach maturity and eggs are produced (fig. 8.3). One approach is to plant a cover crop that is a host to the target nematodes in the fall, while soil temperatures are still adequate for nematode activity (that is, above 60°F). Nematodes enter the plant roots and establish feeding sites. They develop slowly, but their development ceases when the soil temperature drops below the activity threshold. The challenge is to terminate the crop and remove the roots from the soil in the spring before the soil temperature rises above the threshold and nematodes reach the egg-producing stage. If the cover crop is planted early, there will be more nematode activity and penetration into roots, but, depending on the rate of nematode development in relation to soil temperature, it may necessary to terminate the cover crop in the fall. The danger in this tactic is that if the time for crop termination is missed, the nematode life cycle will be completed and there will be new eggs in the soil. Although the method is theoretically appealing, there are only a few examples of it being put into successful practice. Another way that cover crops can act as trap crops for nematodes is to stimulate them to enter roots and establish feeding sites, but prevent them from completing their life cycle. An example is the use of oilseed radish as a trap crop for cyst nematode in fields used for cole crops. This strategy has proved successful in Europe and elsewhere in the United States (Koch et al. 1999), but it has been less effective when tested in California (Gardner and Caswell-Chen 1993).

Mechanism 4: The activity of organisms in the soil food web is increased. This mechanism is important with both Strategy 1 and Strategy 2. Organisms in the soil depend on carbon supplied by plants. Growing cover crops helps mitigate the boom and bust cycle of carbon supply in annual crop agriculture. The typical soil disturbance and fallow periods between crops are both physically and nutritionally disruptive to soil organisms. This disruptive cycle can promote opportunistic plant-feeding species at the entry level of the soil food web and deplete the soil of longer-lived organisms that may be involved in regulation of the pest species but which are sensitive to disturbance (Ferris 2004; Jackson et al. 2004). A very important attribute of cover crops in relation to the soil food web, and its nematode components, is that cover crops are a source of carbon and energy for the soil organisms at times when the soil might otherwise be fallow.

Nematode Host Status of Potential Cover Crops

A search of the world literature and a compilation of local experience provide a summary of current knowledge of the host status of potential cover crops (table 8.2). In some cases, populations of the same nematode species from different regions differ in their ability to reproduce on a cover crop; in other cases differences exist among cultivars in their host status for the same nematode species. Table 8.2 illustrates the need for identification of the nematode species before making a cover crop selection. For example, the host status of cover crops in vegetable production systems differs for the several nematode species that are commonly referred to as root-knot nematode. In the many cases in which the host status of a cover crop has not been tested for a species of nematode, it would be wise to employ Strategy 2, in which the cover is grown when soil temperatures are too low for activity of most plant-feeding nematodes.

Nematodes and the Soil Ecosystem

The bodies of nematodes that feed on bacteria have a higher carbon-to-nitrogen (C:N) ratio than their bacterial prey. Consequently, when nematodes feed on bacteria, they take in more nitrogen than they need. The excess nitrogen is excreted in the form of ammonia, which is available to plants as a nitrogen source. In the presence of a bacterial-feeding nematode community, the nitrogen mineralized and available to plants is increased by 15% or more (Ferris et al. 1997, 2004). Further, nematodes do not digest all the bacteria they consume. As they move through the soil, bacteria that survive passage through a nematode's digestive tract are defecated and exposed to unexploited organic matter in the soil. Consequently, there is more food available to the bacterial community, which increases and provides more food for the nematode community (Fu et al. 2005). This phenomenon is sometimes referred to as nematode farming. When fall cover crops were planted in September rather than November, activity in the soil ecosystem has been shown to increase due to the presence of a carbon source and the elevated soil temperatures (Ferris et al. 2004). Consequently, population levels of bacterial-feeding nematodes increased and were at high levels in the spring when cover crops were incorporated. Nutrients that would have been immobilized in the biomass of bacterial decomposers were mineralized and made available for plant uptake by the feeding activities of the bacterial-feeding nematodes.

Cover crops supply carbon to the soil, promoting a healthy soil food web and populations of predators that can help regulate populations of plant-feeding nematodes. Maintenance of the carbon and energy supply into soil by continuous presence of a photosynthesizing plant, preferably coupled with minimal soil disturbance, is beneficial in the conservation and amplification of the soil food web and its services. Further, the continuous resource supply supports a diversity and abundance of organisms that may disappear with periodic cropping. In ecological terms, greater diversity means greater connectivity and consequently greater stability in the structure and services of a multidimensional food web.

References

Adewusi, S. R. A. 1990. Turnover of dhurrin in green sorghum seedlings. Plant Physiology 94:1219–1224.

Daulton, R. A. C., and R. F. Curtis. 1963. The effects of *Tagetes* sp. on *Meloidogyne javanica* in Southern Rhodesia. Nematologica 9:357–362.

Donkin, S. G., M. A. Eiteman, and P. L. Williams. 1995. Toxicity of glucosinolates and their enzymatic decomposition products to *Caenorhabditis elegans*. Journal of Nematology 27:258–262.

Ferris, H. 1999–2011. NEMAPLEX Nematode-Plant Expert Information System. http://plpnemweb.ucdavis.edu/nemaplex.

———. 2004. Nematodes and the soil food web: Understanding healthy soils. In M. S. Hoddle, ed., Proceedings of the California Conference on Biological Control IV. 1–4. University of California Berkeley Center for Biological Control Web site, http://www.cnr.berkeley.edu/biocon/CCBC%20IV%20Proceedings%202004.pdf.

Ferris, H., and E. P. Caswell-Chen. 1997. Nemabase. UC IPM Web site, http://www.ipm.ucdavis.edu/NEMABASE/.

Ferris, H., C. E. Castro, E. P. Caswell, B. A. Jaffee, et al. 1992. Biological approaches to the management of plant-parasitic nematodes. In J. P. Madden, ed., Beyond pesticides: Biological approaches to pest management in California. Berkeley: University of California Press. 68–101.

Ferris, H., R. C. Venette, and S. S. Lau. 1997. Population energetics of bacterial-feeding nematodes: Carbon and nitrogen budgets. Soil Biology and Biochemistry 29:1183–1194.

Ferris, H., R. C. Venette, and K. M. Scow. 2004. Soil management to enhance bacterivore and fungivore nematode populations and their nitrogen mineralization function. Applied Soil Ecology 24:19–35.

Fu, S., H. Ferris, D. Brown, and R. Plant. 2005. Does the positive feedback effect of nematodes on the biomass and activity of their bacteria prey vary with nematode species and population size? Soil Biology and Biochemistry 37:1979–1987.

Gardner, J., and E. P. Caswell-Chen. 1993. Penetration, development, and reproduction of *Heterodera schachtii* on *Fagopyrum esculentum*, *Phacelia tanacetifolia*, *Raphanus sativus*, *Sinapis alba*, and *Brassica oleracea*. Journal of Nematology 25:695–702.

Gommers, F. J., and J. Bakker. 1988. Mode of action of alpha-terthienyl and related compounds may explain the suppressant effects of *Tagetes* species on populations of free living endoparasitic plant nematodes. In J. Lam et al., ed., Chemistry and biology of naturally occurring acetylenes and related compounds (NOARC). Vol. 7, Bioactive molecules. Amsterdam: Elsevier. 61–69.

Jackson, L. E., H. Ferris, S. R. Temple, K. B. Koffler, and H. Minoshima. 2004. Cover crops, tillage and soil food webs. Sustainable Agriculture Farming Systems Project Newsletter 4(2–3): 1–3.

Koch, D. W., F. A. Gray, and J. R. Gill. 1999. Ten steps to successful trap crop use in the Big Horn Basin. Laramie: University of Wyoming Cooperative Extension Publication B1072.

Timmermans, B. G. H. 2005. *Solanum sisymbriifolium* (Lam.): A trap crop for potato cyst nematodes. Wageningen University Dissertation 3874.

Wang, K.-H., B. S. Sipes, and D. P. Schmitt. 2001. Suppression of *Rotylenchulus reniformis* by *Crotalaria juncea*, *Brassica napus*, and *Tagetes erecta*. Nematropica 31:237–251.

Zasada, I. A., and H. Ferris. 2003. Sensitivity of *Meloidogyne javanica* and *Tylenchulus semipenetrans* to isothiocyanates in laboratory assays. Phytopathology 93:747–750.

———. 2004. Nematode suppression with brassicaceous amendments: Application based upon glucosinolate profiles. Soil Biology and Biochemistry 36:1017–1024.

Zasada, I. A., S. L. Meyer, and C. Rice. 2005. Re-thinking a rye cover crop as a plant-parasitic nematode management tool. Journal of Nematology 37:405.

9 *Robert L. Bugg, William H. Settle, William E. Chaney, and Oleg Daugovish*

Arthropods

This chapter presents an overview of arthropod pest problems that can be affected by cover cropping, practical interventions for minimizing these problems, and a discussion of some of the scientific studies that inform these interventions.

Arthropod pests of California vegetable crops include cabbage aphid *(Brevicoryne brassicae)*, melon aphid (*Aphis gossypii)*, green peach aphid *(Myzus persicae)*, lettuce aphid *(Nasonovia ribisnigri)*, consperse stink bug *(Euschistus conspersus)*, armyworms *(Spodoptera* spp.), cabbage looper *(Trichoplusia ni)*, diamondback moth *(Plutella xylostella)*, false chinch bugs (*Nysius* spp.), common flea beetle *(Epitrix hirtipennis)*, flower thrips (*Frankliniella* spp.), garden symphylan *(Scutigerella immaculata)*, springtails (Collembola), tarnished plant bug *(Lygus hesperus)*, leafminers (*Liriomyza* spp.), and cucumber beetles *(Diabrotica undecimpunctata* and *Acalymma vittata)*.

Several of these pests, given the opportunity, may also feed and reproduce on cover crops, raising the possibility of buildup, dispersal, and damage to adjoining or succeeding vegetable crops. However, there are practical limits on this actually occurring. Specifically, in California vegetable farming systems, fall-sown winter annual cover crops are the dominant types of cover crops grown, and these typically are not allowed to have prolonged flowering, let alone go to seed. Instead, they are usually disked down or mowed before or at peak flower, typically in March or April. Viewed in combination with the normal seasonal developmental patterns of the arthropods in question, this means that relatively few arthropods, whether pest or beneficial, will be able to produce a new generation before the crop is destroyed. This narrows the key pest problems associated with cover crops to aphids, flea beetles, flower thrips, garden symphylan, root maggot, and springtails. Nevertheless, some concerns remain, especially if cover crops are allowed to grow to late flowering or seed set. Below are profiles of several pests that have documented or potential connections to cover crops.

Aphids

Aphids present a special challenge in vegetable production because their asexual reproduction and reproductive potential enable explosive population growth. These features allow one winged female to produce a substantial infestation in a short time. In the Salinas Valley and neighboring areas, cabbage aphid and green peach aphid are important pests in cole crops, and both may be abundant on mustard family cover crops. Bean aphid *(Aphis fabae)* (fig. 9.1) is often abundant on bell bean and is an important vector of viral pathogens to various crops.

Figure 9.1 Bean aphid *(Aphis fabae)*. J. K. CLARK

Figure 9.2. Lettuce aphid *(Nasonovia ribisnigri)*. W. CHANEY

Lettuce aphid *(Nasonovia ribisnigri)* (fig. 9.2), a pest of lettuce that was introduced from Europe during the late 1990s, has proven particularly difficult to control (Parker et al. 2002). It attacks all lettuce varieties but does not occur on cover crops. Lettuce aphid produces asexually reproducing alate (winged) females called viviparae that colonize lettuce at any stage after emergence. Growers usually tolerate this aphid until the plants are thinned about 20 to 30 days after planting. If natural enemies fail to suppress these early colonies, the aphids are treated with insecticide because this insect lives in the heart of the plant, which is protected from most beneficial insects by new leaves that pack around it. Damage may result from feeding, but the primary damage is from contamination of harvested portions of the lettuce with live aphids, cast skins (exuviae), and honeydew.

Data collected by William E. Chaney and colleagues on the Central Coast indicate that parasites and lacewings are not important in the control of this aphid. However, other predators are important, such as aphidophagous hoverflies (Diptera: Syrphidae), which often effectively control lettuce aphid by the time of harvest.

Predator populations in general are enhanced through the use of nectar-bearing plants such as sweet alyssum *(Lobularia maritima)* (Chaney 1998). This low-growing herbaceous plant serves as a nectar source for adult hoverflies and is now commonly used for in-field insectaries in commercial romaine lettuce operations on the Central Coast. Data from California suggest that insectary cover crops could enhance biological control of lettuce aphid (Smith and Chaney 2007).

Melon aphid *(Aphis gossypii)* (fig. 9.3), also known as cotton aphid, is a pest of cucurbit vegetable crops such as squash and melons as well as of cotton. This aphid is capable of explosive population growth, especially during cool periods of the summer or following irrigation (Wilhoit and Rosenheim 1993). Water stress in the host plant inhibits aphid population growth, especially by the large, dark variant of the melon aphid (Wilhoit and Rosenheim 1993). High levels of nitrogen fertilization may also predispose for population increases (Cisneros and Godfrey 2001).

Certain natural enemies are potentially important in biological control of melon aphid. A small wasp, *Lysiphlebus testaceipes* (Hymenoptera: Aphidiidae) (fig. 9.4), is a major parasite (Colfer and Rosenheim 2001). In the San Joaquin Valley, this parasite disperses from maturing wheat in huge numbers during the spring and is probably an important control in cotton during late spring and early summer; it apparently develops well on the large, dark variant of the aphid, but less well on the small, green form (J. Rosenheim, pers. comm.).

The effects of various predators are not always additive in terms of improved biological control. For example, based on data from cage studies, common green lacewing *(Chrysoperla carnea)* (fig. 9.5) is a viable control agent by itself, but intraguild predation on *C. carnea* larvae by the assassin bug *Zelus renardi* and other predatory true bugs (Hemiptera) interferes with this potential biological control (Rosenheim and Wilhoit 1993; Rosenheim et al. 1993, 1999). In Texas, researchers found that relay intercropping with sorghum was useful in building up predators that attack this aphid on cotton and that the predators moved from sorghum to the adjoining cotton (Parajulee et al. 1997; Parajulee and Slosser 1999). These researchers did not address impacts on aphid parasites. It remains to be seen how these results would translate to management of the pest in cucurbit vegetables.

Figure 9.3. Melon aphid *(Aphis gossypii)*, also known as cotton aphid. J. K. Clark

Figure 9.4. A parasitic *Lysiphlebus* spp. wasp (Hymenoptera: Aphidiidae). J. K. Clark

Figure 9.5. Common Green lacewing *(Chrysoperla carnea)* adult. J. K. Clark

Figure 9.6. Consperse stink bug *(Euschistus conspersus)*. J. K. Clark

Other Arthropods

Consperse stink bug *(Euschistus conspersus)* (fig. 9.6) has two generations per year and overwinters as an adult in leaf litter, especially that of blackberry (*Rubus* spp.) in riparian areas (Ehler et al. 2003). Adults disperse in mid to late March. The first generation develops on spring-flowering weeds such as common mustard *(Brassica kaber)*, black mustard *(Brassica nigra)*, wild radish *(Raphanus sativus)*, and cheeseweed *(Malva parviflora)*, finishing development by early June. This raises the possibility of buildup on mustard family cover crops if prolonged spring flowering is permitted. In late May and early June, the first-generation adults disperse to new hosts, such as tomato. In the Sacramento Valley, processing tomato is among the few hosts during most of the summer. Adults or nymphs feeding on green fruit create white, corky lesions under the skin in ripe fruit. In early fall, second-generation adults enter reproductive diapause and fly to overwintering sites.

Garden symphylan *(Scutigerella immaculata)* (fig. 9.7) is of increasing concern to California vegetable growers, in part because registration is being withdrawn for some commonly used soil insecticides. A small (< 10 mm) creature, garden symphylan is not an insect but a member of a separate class, Symphyla. Garden symphylan somewhat resembles the true centipedes in both shape and the manner in which they move, but garden symphylan is white. Garden symphylan feeds on the roots of a wide range of plants, including strawberries, beans, corn, peas, potatoes, parsley, beets, celery, and lettuce (Simigrai and Berry 1974). It also feeds on yeasts, bacteria, fungi, and dead plant and animal material (Edwards 1961). Damage to young plants may result in death or severe stunting, while damage to the roots of older plants provides entry for bacterial and fungal plant pathogens.

Figure 9.7. Garden symphylan *(Scutigerella immaculata)*. J. K. Clark

One early study suggested that while symphylan populations are probably able to survive at a maintenance level by feeding either on directly organic matter or on various saprophytic fungi attacking decaying organic matter, they are not able to reproduce without the addition of fresh plant materials (Shanks 1966). The positive correlation of organic matter and garden symphylans populations may instead be related more to the improved soil physical structure, which would allow them to move more rapidly and extensively through the soil (J. Umble, pers. comm.).

Because symphylans cannot make their own burrows through the soil, they depend on pores and passageways produced by roots, earthworms, and general good soil tilth. This clearly raises an important set of trade-offs. A limited study in the Willamette Valley of Oregon suggested that garden symphylan is more abundant under reduced tillage than under conventional tillage; more abundant following white mustard (*Sinapis alba* 'Monida') than cereal cover crops (spring barley, winter cereal rye, or spring oat); and more abundant following spring barley than spring oat (Peachey et al. 2002).

Root maggots such as the seedcorn maggot *(Delia platura)* (fig. 9.8), attack seedlings of a wide range of vegetable crops and can devastate young stands (Hammond and Cooper 1993; Brust et al. 1997). Larvae can develop on decomposing cover crops and switch to vegetable seedlings. Damage is especially severe in cool weather, because cover crop residue decomposition and seedling development are slowed. If decomposition is prompt and seedling growth rapid, larval development may not occur and seedlings may outgrow injuries. Some vegetable growers recommend flail-chopping cover crops prior to incorporating (I. Morales, American Farms, Soledad, pers. comm.), because finely divided cover crop residue enables more rapid breakdown. This practice appears to be based on sound biological principles.

Tarnished plant bug *(Lygus hesperus)* (fig. 9.9) adults are highly mobile, and the species is famous for building up on natural vegetation or favorable crops, then dispersing to and damaging crops such as cole crops, celery, lettuce, strawberries, and cotton. Adults and nymphs feed on nectar, small insects, and developing seeds. Damage to the latter may result in distorted or aborted fruits (e.g., cotton, strawberries). Harvest of tomato or alfalfa may provoke wholesale flights of adults and their dispersal to adjacent crops (P. Goodell, pers. comm.). Damage to strawberries occurs when feeding on the achenes results in deformed or cat-faced fruit. Damage to celery and lettuce occurs through leaf feeding and results in spots or lesions. Feeding on seedling growing points in cole crops may result in plants that do not produce a flower head, often called blind or blank plants.

San Joaquin Valley cotton growers enrolled in the Sustainable Cotton Project practice strip-mowing of alfalfa to circumvent the mass movements of *Lygus hesperus* that would damage adjoining cotton (M. Gibbs, pers. comm.). In addition, some of these growers have border plantings of annual insectary crops such as buckwheat, cowpea, sorghum, and sunflower to sustain natural enemies of pests and arrest the movement of *L. hesperus.* There are no definitive data on how such approaches work in California vegetable farming systems.

Biological Control

Biological control by pathogens, parasites, and predators assists in controlling arthropod pests of vegetables. Cover crops may harbor arthropod pests and the predators and parasites that are important in biological control. For this and other reasons, trade-offs may be associated with

Figure 9.8. Seedcorn maggot *(Delia platura).* J. K. Clark

Figure 9.9. Tarnished plant bug *(Lygus hesperus).* J. K. Clark

Figure 9.10. Bigeyed bug (*Geocoris* sp.). J. K. Clark

Figure 9.11. Ground beetle (Bembidion quadrimaculatum ssp. dubitans). Adam Martinez, OSAC Specimen #000086748 ©2008 Oregon State Arthropod Collection, OSU, Corvallis Oregon

cover crop selection and management. Cover crop residue can provide a food base for springtails, soil-dwelling mites, and their predators. Some decomposers are facultative herbivores, so they may become crop pests when they exhaust decomposing matter. By contrast, predators that attack decomposers may shift to preying on crop pests, assisting with biological control. Parasites and predators may also occur on living cover crops, feeding on alternate prey or hosts (e.g., aphids or western flower thrips), pollen (e.g., from grasses, legumes, or mustards), or nectar from flowers (e.g., buckwheat) or extrafloral nectaries (e.g., bell bean, common vetch, or cowpea). Table 9.1 presents some commonly used cover crops, the alternative foods they provide, and the natural enemies attracted.

Tamaki and colleagues explored the roles of different predators in field and vegetable crops, emphasizing that different predators may act in combination or sequence and may be important at different stages of pest population growth (Tamaki et al. 1981). For example, generalist predators that occur on the soil surface, such as bigeyed bugs (*Geocoris* spp.) (fig. 9.10), ground beetles (e.g., *Bembidion* spp.) (fig. 9.11), and dwarf spiders (Linyphiidae: Eriogoninae) (fig. 9.12), may be important at low densities of aphid pests. More specialized predators, such as aphid-feeding hoverflies (Syrphidae) are important at higher aphid densities (Tamaki 1972, 1981; Tamaki and Weeks 1972; Tamaki et al. 1981). Other generalist predators that are probably important are brown lacewings (Hemerobiidae) (fig. 9.13), damsel bugs (*Nabis* spp.) (fig. 9.14), green lacewings (Chrysopidae), and minute pirate bug *(Orius tristicolor)* (fig. 9.15).

Beneficial arthropods may colonize vegetable fields

Figure 9.12. Dwarf spider (Linyphiidae: Eriogoninae). J. K. CLARK

Figure 9.13. Brown lacewing *(Hemerobius pacificus)* adult. J. K. CLARK

Table 9.1. Cover crops commonly grown in California vegetable farming systems, alternative foods provided, and arthropod natural enemies attracted

Common name	Scientific name	Alternative foods provided	Arthropod natural enemies attracted	Reference
bell bean	*Vicia faba*	bean aphid *(Aphis fabae)*	lady beetles	Bugg and Ellis 1990
		extrafloral nectar	parasitic wasps, ants	Bugg et al. 1989; Bugg and Ellis 1990; Mondor and Addicott 2003
buckwheat	*Fagopyrum esculentum*	western flower thrips *(Frankliniella occidentalis)* floral nectar	hoverflies, tachinid flies, parasitic wasps, predatory wasps, minute pirate bug	Bugg and Dutcher 1989; Bugg and Ellis 1990
cereal grains: barley, cereal rye, oats, wheat	*Hordeum vulgare, Secale cereale, Avena sativa, Triticum aestivum*	bird cherry oat aphid *(Rhopalosiphum padi)*, corn leaf aphid *(Rhopalosiphum maidis)*, English grain aphid *(Sitobion avenae)*, Russian wheat aphid *(Diuraphis noxia)*	lady beetles	Bugg et al. 1990
common vetch	*Vicia sativa*	cowpea aphid *(Aphis craccivora)*	lady beetles	Bugg et al. 1990
		extrafloral nectar	parasitic wasps, ants	DeBiseau et al. 1997
cowpea (blackeyed pea)	*Vigna unguiculata* ssp. *unguiculata*	cowpea aphid *(Aphis craccivora)* extrafloral nectar	lady beetles, hoverflies, ants parasitic wasps, ants	Bugg and Dutcher 1989 Bugg and Dutcher 1989
mustard family plants (Brassicaceae)	*Brassica* spp.	cabbage aphid *(Brevicoryne brassicae)*, turnip aphid *(Hyadaphis erysimi)*	lady beetles, hoverflies	Bugg et al. 1990
		floral nectar and pollen	hoverflies, green lacewings	Daugovish and Oevering 2003
		western flower thrips *(Frankliniella occidentalis)*	minute pirate bug, predatory thrips	
sorghum, sorghum-sudangrass	*Sorghum bicolor*	greenbug *(Schizaphis graminum)*, corn leaf aphid *(Rhopalosiphum maidis)*	lady beetles	Bugg and Dutcher 1989; Bugg and Ellis 1990

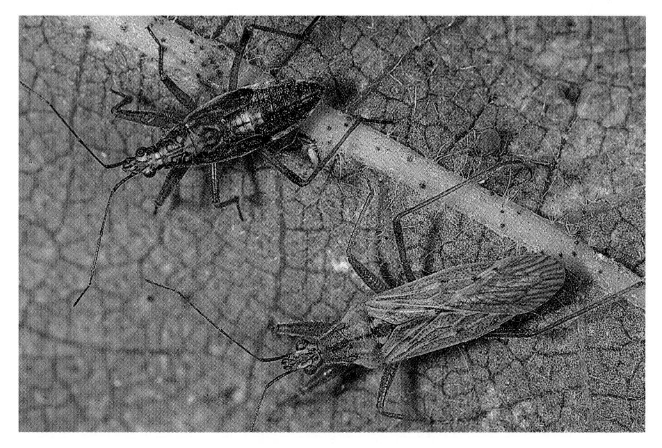

Figure 9.14. Damsel bug (*Nabis* sp.). J. K. Clark

Figure 9.15. Minute pirate bug *(Orius tristicolor)* adult. J. K. Clark

Figure 9.16 Banded-wing thrips (*Aeolothrips* sp.). J. K. Clark

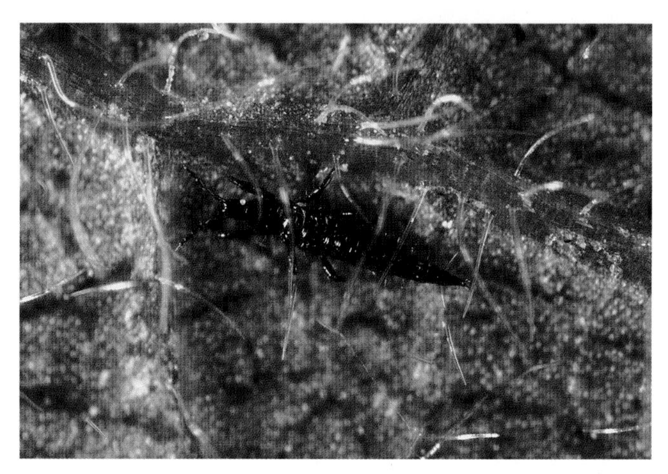

Figure 9.17. Black hunter thrips *(Leptothrips mali)* larva. J. K. Clark

by dispersing from overwintering habitat in wildlands, hedgerows, weedy areas, or agricultural fields. For example, bigeyed bugs are believed to overwinter amid field-side weeds and alfalfa fields and fly to vegetable fields in early spring. By contrast, immature dwarf spiders spin silk and are wafted by the wind, thereby "ballooning" in several successive events to colonize new sites (Weyman et al. 2002; Thomas et al. 2003). Subsequent dispersal between fields may also occur during the growing season. Regardless of their mode of colonizing, beneficial arthropods require resources once they arrive in vegetable fields, and fields of tilled soil and tiny seedling crops may not, by themselves, suffice. Many predatory and parasitic arthropods feed not only on arthropod pests but also on nectar, pollen, and on alternate hosts and prey that may be afforded by non-crop plants. Such plants are termed insectary plants.

In reduced-tillage systems, the bigeyed bug *Geocoris punctipes*, a generalist predator, disperses from living annual clovers to adjoining cotton plants; dead, standing cover crop does not provide such a benefit (Tillman et al. 2004). This represents a form of relay intercropping, with partial overlap in space and time between cover crop and cash crop. Such systems are also used for vegetable farming systems in the southeastern United States with similar results (Bugg et al. 1991), but they are not yet perfected for California.

Beds of sweet alyssum *(Lobularia maritima)* grown as insectary intercrops are commonly used on the Central Coast to provide nectar and alternative prey to aphid-eating hoverflies and other natural enemies of agricultural pests. A complex of entomophagous insects was found in Salinas Valley lettuce fields, and preliminary data suggested that their densities were positively affected by sweet alyssum (Chaney 1998, Bugg et al. 2008).

In Ventura County, California, a 4-year study was conducted of 17 cover crops planted annually in December and terminated in May (mid-flowering) (Daugovish and Oevering 2003). There was weekly sampling to assess arthropod population dynamics in the plant canopy. The study found that

- the greatest densities and diversity in the four sampled groups of natural enemies (pirate bugs, lady beetles, predatory thrips, and miscellaneous parasitoids) were found in vetches, followed by other legumes and yellow mustard
- cereals, bromes, and Indian mustard had low

diversity and low densities of natural enemies, with the exception of a temporary increase in predatory thrips during flowering likely associated with increase in prey (western flower thrips)

- densities of minute pirate bug increased with onset of warmer temperatures in spring in legumes
- all cover crops (especially bell bean, an aphid source) harbored lady beetles
- aphids, flower thrips, and mites were common in legumes
- aphids and thrips were common in cereals
- flea beetles were present at early vegetative stages of mustards
- green lacewings and rove beetles (Staphylinidae) occasionally occurred in all cover crops
- low densities of spiders were present in all cover crops
- in 2 out of 4 years, diamondback moth occurred in mustards
- stinkbugs were commonly found but always in low densities in mustards in all years

Based on these observations, Daugovish and Oevering (2003) recommended avoiding seeding or transplanting brassica vegetables near or immediately following mustard cover crops, to avoid damage from flea beetles and diamondback moth.

Minute pirate bug *(Orius tristicolor)*, predatory thrips (banded-wing thrips *(Aeolothrips* spp.) (fig. 9.16), and black hunter thrips (*Leptothrips mali*) (fig. 9.17), lady beetles (e.g., *Hippodamia convergens*) (fig. 9.18), and miscellaneous parasitoids occurred at high diversity and moderate densities on yellow mustard, low diversity and low densities on oriental mustard, and moderate diversity and high, but not consistently high, densities on bell bean. Dwarf spiders (Araneae: Linyphiidae: Eriogoninae) (Bilde and Toft 2001) prey on cereal aphids and are expected to build up in and disperse from cover crops that include cereal grains. Dwarf spiders are believed to be important in biological control in vegetable crops (Letourneau 1990).

Many entomologists believe that alfalfa fields are especially good habitat for a wide range of beneficial arthropods that disperse readily to nearby vegetable fields (Rackickas and Watson 1974). Hedgerows, windbreaks, and wildlands may also provide important habitat. Small-scale dispersal of arthropods has been studied within strip-mowed stands of alfalfa (Hossain et al. 2002), indicating similar short-distance movement of predators and pests from mowed into unmowed strips, which afford more hospitable habitat. Simultaneous intercrops of perennial legumes, however, do not necessarily result in higher parasitism of aphids on broccoli (Costello and Altieri 1994). These kinds of intercrops are used by some small-scale growers in California.

A landscape-scale study of processing tomatoes in the Central Valley found that winter cover cropping was statistically associated with elevated densities of common flea beetle *(Epitrix hirtipennis)* and three parasitic wasps that attack vegetable leafminer (Letourneau and Goldstein 2001). Movement by arthropods varies not only among species but also seasonally (Corbett 1998). Rubidium labeling has been used to demonstrate that predatory and parasitic insects feed on nectar in insectary cover crops and hedgerows and may then disperse widely into adjoining agricultural fields and orchards (Long et al. 1998).

Conclusion

Cover crops can provide a useful tool for managing insect pests in vegetable crops. However, they can have unintended negative consequences as well. Cover crops are often grown at separate times from vegetable crops and may have little impact on their arthropod populations. However, standing cover crops can provide habitat for beneficial as well as pest insects and may affect surrounding vegetable crops. It is therefore critical that the impact of cover crops on insects be considered when choosing cover crop species. In addition, cover crop residue can impact soilborne insects and also must be considered in both the choice of cover crop species and their management prior to planting vegetable crops.

References

Bachi, O., A. Ploeg, G. Walker, and M. E. McGiffen. 2009. Interdisciplinary vegetable pest management potentials of selected cover crops. HortScience 44(4): 1112–1113.

Figure 9.18. Lady beetle *(Hippodamia convergens)* adults. J. K. CLARK

Bilde, T., and S. Toft. 2001. The value of three cereal aphid species as food for a generalist predator. Physiological Entomology 26:58–68.

Brust, G. E., R. E. Foster, and W. Buhler. 1997. Effect of rye incorporation, planting date, and soil temperature on damage to muskmelon transplants by seedcorn maggot (Diptera: Anthomyiidae). Environmental Entomology 26:1323–1326.

Bugg, R. L., and J. D. Dutcher. 1989. Warm-season cover crops for pecan orchards: Horticultural and entomological implications. Biological Agriculture and Horticulture 6:123–148.

Bugg, R. L., and R. T. Ellis. 1990. Insects associated with cover crops in Massachusetts. Biological Agriculture and Horticulture 7:47–68.

Bugg, R. L., R. T. Ellis, and R. W. Carlson. 1989. Ichneumonidae (Hymenoptera) using extrafloral nectar of faba bean (*Vicia faba* L., Fabaceae) in Massachusetts. Biological Agriculture and Horticulture 6:107–114.

Bugg, R. L., S. C. Phatak, and J. D. Dutcher. 1990. Insects associated with cool-season cover crops in southern Georgia: Implications for pest control in the truck-farm and pecan agroecosystems. Biological Agriculture and Horticulture 7:17–45.

Bugg, R. L., F. L. Wäckers, K. E. Brunson, J. D. Dutcher, and S. C. Phatak. 1991. Cool-season cover crops relay intercropped with cantaloupe: Influence on a generalist predator, *Geocoris punctipes* (Hemiptera: Lygaeidae). Journal of Economic Entomology 84:408–416.

Bugg, R. L., R. G. Colfer, W. E. Chaney, H. A. Smith, and J. Cannon. 2008. Flower flies (Syrphidae) and other biological control agents for aphids in vegetable crops. Oakland: University of California Agriculture and Natural Resources Publication 8285. ANR Communication Services Web site, http://anrcatalog.ucdavis.edu/Items/8285.aspx.

Chaney, W. E. 1998. Biological control of aphids in lettuce using in-field insectaries. In C. H. Pickett and R. L. Bugg, eds., Enhancing biological control: Habitat management to promote natural enemies of arthropod pests. Berkeley: University of California Press. 73–83.

Cisneros, J. J., and L. D. Godfrey. 2001. Midseason pest status of the cotton aphid (Homoptera: Aphididae) in California cotton: Is nitrogen a key factor? Environmental Entomology 30:501–510.

Colfer, R. G., and J. A. Rosenheim. 2001. Predation on immature parasitoids and its impact on aphid suppression. Oecologia 126:292–304.

Corbett, A. 1998. The importance of movement in the response of natural enemies to habitat manipulation. In C. H. Pickett and R. L. Bugg, eds., Enhancing biological control: Habitat management to promote natural enemies of arthropod pests. Berkeley: University of California Press. 25–48.

Costello, M. J., and M. A. Altieri. 1994. Living mulches suppress aphids in broccoli. California Agriculture 48:24–28.

Daugovish, O., and P. E. Oevering. 2003. Cover crop evaluation for biomass production, weed competitiveness and as hosts for beneficial insects. Proceedings of the American Society of Agronomy, Annual Meeting, Denver, Colorado.

Edwards, C. A. 1961. The ecology of Symphyla, Part III: Factors controlling soil distributions. Entomologia Experimentalis et Applicata 4:239–256.

Ehler, L. E., C. G. Pease, and R. F. Long. 2003. Farmscape ecology of a native stink bug in the Sacramento Valley. Fremontia 31(1).

Hammond, R. B., and R. L. Cooper. 1993. Interaction of planting times following the incorporation of a living, green cover crop and control measures on seedcorn maggot populations in soybean. Crop Protection 12:539–543.

Hossain, Z., G. M. Gurr, S. D. Wratten, and A. Raman. 2002. Habitat manipulation in lucerne (*Medicago sativa* L.): Arthropod population dynamics in harvested and "refuge" crop strips. Journal of Applied Ecology 39:445–454.

Letourneau, D. K. 1990. Mechanisms of predator accumulation in a mixed crop system. Ecological Entomology 15(1): 63–69.

Letourneau, D. K., and B. Goldstein. 2001. Pest damage and arthropod community structure in organic vs. conventional tomato production in California. Journal of Applied Ecology 38(3): 557–570.

Long, R. F., A. Corbett, C. Lamb, C. Reberg-Horton, J. Chandler, and M. Stimmann. 1998. Movement of beneficial insects from flowering plants to associated crops. California Agriculture 52(5): 23–26.

Mondor, E. B., and J. F. Addicott. 2003. Conspicuous extra-floral nectaries are inducible in *Vicia faba*. Ecology Letters 6:495–497.

Parajulee, M. N., and J. E. Slosser. 1999. Evaluation of potential relay strip crops for predator enhancement in Texas cotton. International Journal of Pest Management 45:275–286.

Parajulee, M. N., R. Montandon, and J. E. Slosser. 1997. Relay intercropping to enhance abundance of insect predators of cotton aphid (*Aphis gossypii* Glover) in Texas cotton. International Journal of Pest Management 43:227–232.

Parker, W. E., R. H. Collier, P. R. Ellis, A. Mead, D. Chandler, J. A. B. Smyth, and G. M. Tatchell. 2002. Matching control options to a pest complex: The integrated pest management of aphids in sequentially-planted crops of outdoor lettuce. Crop Protection 21(3): 235–248.

Peachey, R. E., A. Moldenke, R. D. William, R. Berry, E. Ingham, and E. Groth. 2002. Effect of cover crops and tillage system on symphylan (Symphyla: *Scutigerella immaculata* Newport) and *Pergamasus quiquiliarum* Canestrini (Acari: Mesostigmata) populations, and other soil organisms in agricultural soils. Applied Soil Ecology 21:59–70.

Rackickas, R. J., and T. F. Watson. 1974. Population trends of *Lygus* spp. and selected predators in strip-cut alfalfa. Journal of Applied Ecology 3:781–784.

Rosenheim, J. A., and L. R. Wilhoit. 1993. Predators that eat other predators disrupt biological control of the cotton aphid. California Agriculture 47(5): 7–9.

Rosenheim, J. A., L. R. Wilhoit, and C. A. Armer. 1993. Influence of intraguild predation among generalist insect predators on the suppression of an herbivore population. Oecologia 96:439–449.

Rosenheim, J. A., D. D. Limburg, and R. G. Colfer. 1999. Impact of generalist predators on a biological control agent, *Chrysoperla carnea*: Direct observations. Ecological Applications 9:409–417.

Shanks, C. A. 1966. Factors that affect reproduction of the garden symphylan *Scutigerella immaculata*. Journal of Economic Entomology 59:1403–1406.

Simigrai, M., and R. E. Berry. 1974. Resistance in broccoli to the garden symphylan. Journal of Economic Entomology 67:371–373.

Smith, H. A., and W. E. Chaney. 2007. A survey of syrphid predators of *Nasonovia ribisnigri* in organic lettuce on the Central Coast of California. Journal of Economic Entomology 100:39–48.

Smith, H. A., W. E. Chaney, and T. A. Bensen. 2008. Role of syrphid larvae and other predators in suppressing aphid infestations in organic lettuce on California's Central Coast. Journal of Economic Entomology 101:1526–1532.

Tamaki, G. 1972. The biology of *Geocoris bullatus* inhabiting orchard floors and its impact on *Myzus persicae* on peaches. Environmental Entomology 1:559–565.

———. 1981. Biological control of potato pests. In J. H. Lashomb and R. Casagrande, eds., Advances in potato pest management. Stroudsberg, PA: Hutchison Ross. 178–192.

Tamaki, G., and R. E. Weeks. 1972. Biology and ecology of two predators, *Geocoris pallens* Stål and *G. bullatus* (Say). USDA Agricultural Research Service Technical Bulletin 1446.

Tamaki, G., B. Annis, and M. Weiss. 1981. Response of natural enemies to the green peach aphid in different plant cultures. Environmental Entomology 10:375–378.

Thomas C. F. G., P. Brain, and P. C. Jepson. 2003. Aerial activity of linyphiid spiders: Modelling dispersal distances from meteorology and behaviour. Journal of Applied Ecology 40(5): 912–927.

Tillman G., H. Schomberg, S. Phatak, B. Mullinix, S. Lachnicht, P. Timper, and D. Olson. 2004. Influence of cover crops on insect pests and predators in conservation tillage cotton. Journal of Economic Entomology 97(4): 1217–1232.

Weyman, G. S., K. D. Sunderland, and P. C. A. Jepson. 2002. A review of the evolution and mechanisms of ballooning by spiders inhabiting arable farmland. Ethology, Ecology and Evolution 14(4): 307–326.

Wilhoit, L. R., and J. A. Rosenheim. 1993. The yellow dwarf form of the cotton aphid *Aphis gossypii*. In Proceedings of the Beltwide Cotton Conference, New Orleans. Memphis: National Cotton Council of America. 969–972.

PART 4

Cover Cropping Systems and Economics

10 *Mark Van Horn, Eric Brennan, Oleg Daugovish, and Jeff Mitchell*

Cover Crop Management

The benefits derived from growing a cover crop depend on the type of cover crop grown and how it is managed. This chapter discusses a sequence of cover crop management decisions and practices that are generally applicable throughout California. Cover crop management can vary considerably among the different regions of the state; information about regionally specific practices, such as recommended species, varieties, and planting dates, can be found in chapter 11.

Timing

Cool- and warm-season cover crops can be grown in most vegetable production regions of California, although cool-season cover crops are grown more often. The planting date and total length of growing season can impact several cover crop performance factors, such as biomass production, weed suppression, nitrogen fixation, and nitrogen scavenging. This is particularly true with cool-season legumes and legume-cereal mixtures that often produce poor or weedy stands, or fail completely, when they are planted after recommended dates. Conversely, cool-season grass cover crops that are planted too early in the fall may flower prematurely, reducing cover crop biomass production and possibly producing seed that can germinate and become weeds in subsequent crops.

The range of acceptable planting dates is generally greater in the more moderate coastal regions than in the interior valleys. This range is also fairly broad for warm-season cover crops in most of the state because of the relatively long period of suitable conditions and the relatively short growing season required, although water limitations may restrict the timing of warm-season cover cropping in some areas.

Field Preparation and Planting

Field preparation and planting of cover crops typically follow standard practices used with agronomic cash crops. The main objective of these activities is to place the seed at the proper depth (see table 2.1 in chapter 2) with good soil-to-seed contact to promote uniform germination and emergence. Cover crops are usually more tolerant of marginal planting conditions than are vegetable crops. Field preparation activities for cover crops can range from none, such as when using no-till planters (fig. 10.1), to extensive and may include making furrows for irrigation or drainage of the cover crop or subsequent cash crops.

Most cover crops are planted as solid stands with grain drills that place the seed in rows spaced 6 to 10 inches apart (fig. 10.2). Solid stands also can be achieved by broadcasting the seed and incorporating it with

Figure 10.1. No-till grain drill plants a triticale-pea-vetch cover crop into cotton residue. J. MITCHELL

Figure 10.2. Drilled stand of cereal rye. R. SMITH

implements such as harrows or ring rollers, depending on soil conditions and desired seeding depth (fig. 10.3). Broadcasting produces less-uniform seed distribution, seeding depths, and crop stands and thus requires higher seeding rates. Seeding with a grain drill on beds may require adjusting the drill to achieve the proper seeding depth in all rows or using a special planting configuration. For example, certain rows on a grain drill may be plugged to prevent planting in the furrows. Some cover crops, such as cowpeas, can be planted with bean or corn planters in rows spaced farther apart (e.g., 28 to 40 inches), allowing in-season cultivation for weed control.

Cover crop mixtures (fig. 10.4) are usually planted by using seed mixes that are placed in a single seed hopper of the planter (Figure 10.5). However, drills with two seed hoppers can be used if the seeds of the different species do not stay well mixed in a single hopper. Mixtures of species with seeds of very different sizes are commonly planted at a depth that is a compromise among the optimal depths for the different species.

Cover crops that are integrated into conservation tillage rotations can be established using no-till drills that can seed directly into residues using disk openers, coulters, and hydraulic pressure. Cover crops have been established successfully in California after the harvest of tomatoes, melons, corn, cotton, grain legumes, and wheat with either no preplant tillage or after one or two passes with a minimum-till disk.

Seeding Rate

Table 2.1 (chapter 2) gives recommended cover crop seeding rates for generally favorable conditions. However, the actual optimal seeding rate for a given situation can vary considerably depending on many factors. In general, higher seeding rates are needed as conditions become less favorable or more risky for cover crop germination and growth or when high cover crop densities are needed to achieve a specific objective early in the season. Depending on the situation, seeding rates may be only slightly more than the upper end of the range listed in table 2.1 or more than 1.5 times that rate. Using a higher seeding rate is recommended if

- the germination rate is lower than 90%, the minimum rate assumed in table 2.1 (to compensate for low germination rates, divide the recommended seeding rate by the germination percentage, expressed as a proportion; e.g., if the recommended rate is 100 lb/acre and the germination rate is 80%, plant at 100 ÷ 0.80, or 125 lb/acre)
- the thousand kernel weight of the seed is greater than that listed in table 2.1
- the planting method (e.g., broadcast and harrow) does not consistently place seed at the proper depth or spacing
- the planting date is significantly earlier or later than recommended or there is a risk that the germination date of a nonirrigated cool-season cover crop may be

Fig. 10.3. Broadcasting mustard seed on the soil surface from a small, hydraulically driven seed hopper and shallowly incorporating it with a spike-tooth harrow. R. Smith

Figure 10.4. Legume-cereal cover crop mixture seeded with a grain drill. R. Smith

Figure 10.5. Legume-cereal cover crop seed mix in grain drill hopper. R. Smith

Figure 10.6. Seeding rate and variety affect how quickly drilled rows of cereal rye provide total ground cover to suppress weed growth between seed rows. R. Smith

late because of a delayed onset of the rainy season

- the growing season for a cover crop will be very short, and total production will be directly influenced by cover crop stand density
- denser early-season cover crop stands are desired to be more competitive with weeds
- denser early-season cover crop stands are desired to be more effective in absorbing residual soil nitrogen

Cover crop seed typically accounts for a relatively small part (20 to 30%) of cover cropping costs (see chapter 12), but the seeding rate used can have a large effect on some of the beneficial effects of cover cropping. For example, in a long-term cover crop trial, weed densities were six times higher in organic lettuce that had 5 years of a rye-legume winter cover crop planted at 125 versus 375 pounds per acre (E. Brennan, unpublished data). Because the objectives and conditions of cover cropping can vary greatly, growers may want to experiment with different seeding rates in different areas of a field and carefully observe and record the cover crop's performance throughout the season. This information may assist in decision making in future years. Various seeding rates can easily be achieved by adjusting the seeding rate lever on most grain drills or by planting a section of a field two or more times. Because fields are inherently variable it is better to have a number of small test areas than one large one. Important factors that are typically affected by seeding rate and are relatively easy to monitor include the number of cover crop plants per foot of row, weed density (e.g., number of weeds per square foot) and growth, number of days to canopy closure (fig. 10.6), and cover crop height, flowering date, and degree of lodging.

Seeding Rate Calculators

Growers can use simple online seeding rate calculators such as the one at the Government of Alberta's Agriculture and Rural Development Decision-Making Tools Web site (http://www.agric.gov.ab.ca/app19/calc/crop/otherseedcalculator.jsp) to calibrate their planting equipment and determine how to adjust the seeding rate for a cover crop based on factors such as thousand kernel weight, row spacing, and germination rate. These calculators can help you achieve a target plant density and can provide insight into why cover crops perform differently at various rates. Consider the following example, which has a target density of 45 rye plants per square foot. Assume there is 90% germination and 5% seedling emergence mortality. Using the online calculator, a seeding rate of 81 pounds per acre would be required for seed with a thousand kernel weight of 16 grams, whereas a seeding rate of 111 pounds per acre would be necessary for seed with a thousand kernel weight of 22 grams. Note that seeding rate is independent of row spacing, number of acres planted, and seed price, but the calculator will not work unless a row spacing value is entered.

Seeding rate decisions are more complex in cover crop mixtures than in single-species cover crops because some components (typically legumes) of mixtures are often more sensitive to competition as seeding rate increases. Competition between components of a mixture at a given seeding rate can also be influenced by nutrient availability and ambient temperature. For example, the grass component of a grass-legume mixture is more likely to dominate the mixture in a soil with high residual nitrogen than in a soil with low nitrogen.

Nutrient Management, Fertilization, and Inoculation

Cover crops can improve nutrient management in a cropping system by recycling nutrients, reducing losses from leaching and runoff, and adding nitrogen. Nonlegume cover crops such as cereals and mustards are the most effective scavengers of nitrogen that might otherwise leach from the topsoil and pollute ground or surface waters. Legume cover crops can add nitrogen to a cropping system when *Rhizobia* bacteria present in their roots convert atmospheric nitrogen into a form that plants can use in the process of nitrogen fixation (see chapter 5). Cover crop mixtures of legumes and nonlegumes can combine the nitrogen fixing of the legumes with the nitrogen scavenging of the nonlegumes.

Nitrogen fixation can occur only when the strain of *Rhizobia* present in a legume's roots is an effective strain for that particular species of legume. Therefore, legume cover crop seed should always be inoculated with the appropriate strain of *Rhizobia.* Inoculants can be purchased from seed dealers and other suppliers and stored in a sealed container in a cool location until use. Inoculants should be used prior to their expiration dates; once inoculated, seed should be kept out of direct sunlight and planted as soon as possible. It is also possible to purchase preinoculated seed of many legumes. Legume root nodules that have a pink to red color internally provide evidence of effective nodulation and nitrogen fixation (fig. 10.7). Fertilizing legume cover crops with nitrogen is not recommended because nitrogen fixation is inhibited by available soil nitrogen. However, additions of lime or phosphorous fertilizers may improve nitrogen fixation in some situations, such as on soils that are acidic or have low levels of available phosphorous.

Most vegetable-producing soils in California are highly fertile and can produce vigorous cover crops without fertilizers. Therefore, fertilization of cover crops is generally discouraged. However, if a non-nitrogen nutrient deficiency exists, correcting the deficiency prior to planting a cover crop will improve its performance, and the nutrients taken up by the cover crop will ultimately be returned to the soil when the cover crop decomposes. In addition, nitrogen fertilization of a warm-season nonlegume cover crop such as sudangrass may be warranted when it is grown on a nitrogen-deficient soil and the objective is to add large amounts of organic matter to the soil.

Figure 10.7. Nodulation on roots of bell bean. Note the lower nodule, cut open to show the pink color indicating effective nitrogen fixation. R. SMITH

Irrigation

Irrigation can be an important part of successful cover crop management. Pre- or postplant irrigation of cool-season cover crops can help ensure germination and vigorous cover crop growth early in the fall. If used, sprinkler pipe should be removed from the field soon after irrigation is complete to avoid difficulties in removing it later in the season after significant cover crop growth. In-season irrigation of cool-season cover crops is uncommon but may improve cover crop growth in drier locations and seasons. Irrigation is essential for most warm-season cover crops in California, and most species are more productive when they have more water. However, some species (e.g., sorghum, sudangrass, cowpea) can be productive on deep, fertile soils with minimal irrigation if soil moisture levels are initially high.

End-of-Season Management

At the end of the cover cropping season, growers must make a number of interrelated decisions about the timing and methods used in cover crop termination, field preparation, and planting of the subsequent vegetable crop. These decisions may be influenced by the type and purpose of the cover crop; the cover crop's stage of development, biomass, and nitrogen and fiber content; the soil type and soil moisture content; and prevailing weather patterns. For example, with grass cover crops, which can produce large amounts of slowly decomposing, tough, fibrous residue, a grower may terminate the cover crop long before it reaches full vegetative growth to minimize problems associated with managing such residues (fig. 10.8). Some grass cover crops, particularly *Sorghum* species, produce allelopathic compounds that can damage subsequent cash crops if the compounds persist into the crop planting period. In some situations, early cover crop termination may be necessary to perform field operations needed to prepare the field adequately for a subsequent vegetable crop.

Figure 10.8 The grass residue on the right was more difficult to incorporate into the soil than the legume residue on the left. R. SMITH

Table 10.1. Conversion factors for estimating nitrogen content of selected cover crops

Crop common name	Scientific name	Factor
bell beans	*Vicia faba*	10
sesbania	*Sesbania* spp.	10
sunn hemp	*Crotalaria juncea*	11
blackeyed peas (cowpeas)	*Vigna unguiculata* ssp. *unguiculata*	12
berseem clover	*Trifolium alexandrinum*	13
woolypod vetch	*Vicia villosa* ssp. *varia*	16
purple vetch	*Vicia benghalensis*	16

Conversely, if the objective is to maximize the biomass or nitrogen additions from a cover crop, the crop should be terminated during early to mid bloom because biomass and nitrogen accumulation rates are typically highest just before bloom and decrease dramatically at the onset of flowering.

Estimating the Nitrogen Content of a Legume Cover Crop

When deciding how long to maintain a legume cover crop that is being grown for its nitrogen benefit, it can be helpful to estimate the amount of nitrogen in the standing cover crop using the procedure described below. However, the need to prepare the field and plant a cash crop by a certain date or as the weather or soil moisture conditions permit sometimes overrides all other considerations in determining when to terminate a cover crop.

1. Cut and weigh the fresh aboveground cover crop biomass from an area of 16 square feet (e.g., 4 ft by 4 ft) (fig. 10.9). The plants should be free of dew. Flowers may be present, but seed pods should not be. Remove any weed biomass from the sample before weighing.
2. Use the appropriate conversion factor from table 10.1 and the equation below to estimate the number of pounds of nitrogen per acre contained in the cover crop. For cover crop mixtures, separate the components of the mixture, individually weigh and calculate the nitrogen contribution of each component, and add the individual contributions for the total nitrogen contribution.

$$\text{Fresh weight (lb)} \times \text{Conversion factor} = \text{Nitrogen in cover crop (lb/ac)}$$

3. Repeat this procedure several (e.g., 5 to 10) times over the field to get an idea of the average amount of nitrogen in the cover crop and the amount of variability in the field.

Terminating the Cover Crop

Growers can use a number of methods to terminate a cover crop and prepare and plant the field to the following cash crop. Mowing is important in most strategies because residues are managed more easily and decompose more rapidly when shredded into fine pieces with a mower (fig. 10.10). Flail mowers are preferable to rotary mowers because they cut the cover crop into finer pieces and distribute them more evenly on the soil surface. Usually, the mowed residue should be incorporated as soon as possible. However, this is not always feasible, and in a wet year it may be worthwhile to mow a cover crop to prevent seed set by the cover crop or weeds even if residue incorporation will be delayed until the soil dries sufficiently.

A typical sequence of operations includes mowing, disking two or three times, and making beds. When large amounts of fibrous residues are present, such as is common with grass cover crops, making several additional passes with a disk or moldboard plow may be needed to prepare the field adequately for planting vegetables. In such situations, a spader may be a useful alternative to more conventional implements. Spaders travel at slow speeds (e.g., $^1/_4$ to $^3/_4$ mph) but can incorporate residues thoroughly in a single pass (fig. 10.11). They can be particularly useful for smaller-scale diversified vegetable operations that often have areas that are too small to be disked effectively, regardless of the type and amount of cover crop residue involved. A minimum-till disk can preserve preexisting beds and may be an option for reducing tillage if low or moderate amounts of residue are present. Other minimum-till options for managing cover crops include using various rollers, crimpers, and stalk choppers (fig. 10.12) to convert cover crops into surface mulches that may be followed by strip tillers that till only a narrow band on the bed top for planting or transplanting the vegetable crop. However, for cover crops species that must be killed completely before planting the subsequent crop, rolling, crimping or chopping may not be adequate in all situations.

Figure 10.9. The nitrogen content of this cover crop was estimated in the field using the materials shown and the procedure described in the text. M. VAN HORN

Figure 10.10. Mustard cover crop mowed with a flail mower prior to disking. R. SMITH

Figure 10.11. A spader incorporates standing cover crop of mixed legumes in one pass. M. VAN HORN

Figure 10.12. Front-mounted roller-crimper flattens rye cover crop while rear-mounted seeder plants cotton. J. MITCHELL.

References

Government of Alberta, Agriculture and Rural Development. 2010. Seeding rate calculator Web site, http://www.agric.gov.ab.ca/app19/calc/crop/otherseedcalculator.jsp.

Phillips, D. A., and W. A. Williams. 1987. Range-legume inoculation and nitrogen fixation by root-nodule bacteria. Oakland: University of California Agriculture and Natural Resources Publication 1842.

11 Regional Differences in Cover Crop Management

Given the diversity of the state of California, significant differences exist in cover cropping practices between the Central Valley, desert, and coastal production regions. The sections in this chapter illustrate general differences in management among these regions.

Central Valley Region

Brenna Aegerter, Joe Nuñez, and Mark Van Horn

Background

Commercial vegetable production in the Central Valley is concentrated in the San Joaquin Valley and southern Sacramento Valley. Vegetables produced in this region include fresh market and processing tomatoes, processing garlic and onions, and melons. Potatoes, carrots, and leafy vegetables are also produced, concentrated in the southern San Joaquin Valley. The Westside area of Fresno and Kings Counties is a major producer of lettuce and tomatoes. The Delta region is known for asparagus production, and the Sacramento region is known for growing processing tomatoes. Organic vegetable producers and small-scale growers who produce a wide mix of summer and winter vegetables are scattered throughout much of the Central Valley.

Diverse row crops are grown in the Central Valley, so there is no typical rotation. In general, however, a standard rotation in the northern areas would be tomatoes rotated with small grains or grain corn. In the central part of the Valley, cotton, melons, tomatoes, and leafy vegetables are commonly rotated. At the southern end of the valley, there is a greater diversity of crops, but common rotations are carrots-potatoes-small grains, garlic-carrots-tomatoes, and peppers-carrots-small grains. Throughout the Valley, other rotation crops might include alfalfa, safflower, sugar beets, or other vegetables.

The Central Valley has a Mediterranean climate with hot, dry summers and cool, moist winters. As the valley spans 400 miles from north to south, the many subclimates range from arid to semiarid. More than 80% of the precipitation falls in the winter, with average annual precipitation varying from 17 inches in the Sacramento area to 5.5 inches in the southern San Joaquin Valley. The maximum temperature during the summer may exceed 100°F on several days, and average daily maximums in July range from 93°F (Sacramento) to 99°F (Bakersfield). Where the Sacramento and San Joaquin Rivers meet before they empty into San Francisco Bay, a marine influence moderates temperatures, the most important effect being lower summer nighttime temperatures. Central Valley winters are generally mild, with average high temperatures in the low to mid 50s and low temperatures in the high 30s to low 40s. In the winter, dense tule fog is common.

Cover cropping is practiced on a small percentage of the conventional vegetable acreage in the Central Valley. With the potential for growing vegetable crops year-round, many growers are reluctant to invest the time and money to grow cover crops in lieu of cash crops. Additionally, winter cover crops may interfere with spring planting schedules for warm-season crops.

Many growers realize the benefits of using cover crops, and some try to incorporate cover cropping into their farm operations when it fits the crop planting schedule. Organic growers generally plant cover crops more frequently than do conventional growers in order to supply nitrogen to subsequent crops or improve soil physical properties. Some large-scale farming operations routinely grow cover crops when the ground would normally be fallow.

Cover Crop Production

Winter production

Barley is the most common cover crop grown in the San Joaquin Valley. It is usually planted from October through January and incorporated as a green manure in late winter or early spring. Often a cover crop of barley or oat mixed with peas and vetches is planted. These mixes have the benefits of providing green manure biomass as well as fixing nitrogen by the legumes. The cover crop is typically disked under when the plants reach 18 to 24 inches tall, and sufficient time is allowed for the green manure to break down before a vegetable crop is planted.

Some growers plant grass or legume mix cover crops on shaped tomato beds over the winter. Grass cover crops are killed with herbicide to keep biomass levels on bed tops manageable. Residue on the beds is then tilled prior to planting, or tomatoes are transplanted through the residue. The benefit of this system has been to reduce runoff volume, but the levels of reductions depend on the intensity of storm events and soil moisture levels; a yield benefit may be seen with this system. Some processing tomato acreages are managed with conservation tillage, in which grass cover crops are grown on beds over the winter. The cover crops are rolled with a crimper when the cover crop flowers to kill them, and tomatoes are transplanted into the residue.

Growers wishing to use cover crops as green manures to supply nitrogen to their rotations plant legumes or legume-grass mixtures that are 80 to 90% legumes by weight. Vetches, field peas, and bell beans, grown singly or in mixes, are the most common legumes, while cereals such as oat, rye, or barley are the most common grasses. In the Sacramento Valley, these legumes are planted and germinated by mid-October to allow for sufficient fall growth to ensure good overall productivity and competition against weeds. Such cover crops are typically terminated in March or early April.

Recently there has been an interest in the use of mustards as winter cover crops. Mustards produce a large amount of biomass; when incorporated as a green manure the decomposing residue releases the same chemical that is produced by metam sodium fumigation (see chapter 2). However, the quantity of fumigant produced by the mustard is only a small fraction of that contained in standard rates used in commercial fumigation and therefore is not as effective at reducing soilborne pests. No soilborne disease suppression has been seen in recent mustard trials conducted on processing tomatoes. In spite of the limitations, some growers have been using mustards in areas where fumigation cannot be used because of proximity to homes, schools, or other buildings. Mustards are planted from fall through spring and incorporated 60 to 90 days after planting. The mustard is flail-chopped and immediately incorporated into the soil with a disk. Often, water is applied by sprinkler irrigation following incorporation. Mustards do not produce significant biomass if planted in the warmer times of the year in the Central Valley.

Summer production

Sudangrass and hybrids of sudangrass and sorghum (e.g., Sordan 79) are occasionally used as summer cover crops. These crops may be mowed several times before being incorporated into the soil. Some cultivars produce a biofumigation effect similar to mustards that suppresses soilborne pests. For example, Sordan 79 and Trudan 8 are nonhosts for root-knot nematode but also produce a chemical that acts like a fumigant when incorporated into the soil (see chapter 8). Sudangrass and its hybrids are planted in the spring and summer months and are incorporated when they reach 4 to 6 feet tall. Although grown on relatively few acres, summer legume cover crops, such as cowpea and crotalaria, can be grown to provide moderate amounts of biomass and nitrogen in a relatively short period of 2 to 3 months.

Although cover cropping is not widely practiced in this region, growers understand the benefits of cover crops. However, growing them may not be economically justifiable when other more profitable crops can be grown or when the added expense to grow a cover crop is considered. Despite these limitations, some growers do use cover crops on a regular basis in the San Joaquin and Sacramento Valleys.

Desert Region

José Aguiar and Milton E. McGiffen Jr.

Background

The low desert climate is characterized by annual rainfall of 3 inches or less, long, hot summers with many days over 100°F, and mild winters with daily highs around 70°F and rare frosts. The main vegetable production areas include the Coachella Valley and Palo Verde in Riverside County and the Imperial Valley in Imperial County. In Palo Verde and the Imperial Valley, cotton, alfalfa, and most other field crops are the dominant crops, but a significant acreage of lettuce is also produced. Cover crops are rarely included in the rotations. Coachella Valley vegetable production is intensive, often with two crops per year, including leaf lettuce and other winter vegetables planted from

August through November, and bell peppers and melons are planted from January through February.

Desert soils are low in organic matter, and cover crops are grown primarily to increase organic matter and to reduce wind erosion and dust pollution.

Cover Crop Production

Winter production

Because of the importance of winter vegetables, winter cover crops are uncommon in the desert. In the rare instances that winter cover crops are grown, barley, oat, rye, peas, or vetch are used, and the planting period is November to February.

Summer production

In the desert the land is traditionally fallowed during July and August. These months make an excellent time for planting specific cover crops that grow well under warm conditions. The two most commonly used cover crops are sudangrass and cowpea. Both of these flourish when temperatures exceed 90°F, which is common from April through September in most of the low desert. Both cowpea and sudangrass will produce abundant biomass if grown for at least 6 weeks in the midsummer heat, and for a longer period at other times of the year.

Sudangrass is a C4 grass (well adapted to and grows well in high temperatures) that rapidly produces biomass and can scavenge residual soil nitrate and other nutrients. Cowpea, a legume, can add over 100 pounds of nitrogen per acre to the soil. Cowpeas and sudangrass are typically planted from May to July and are both adaptable to various production systems. Cowpeas are often grown in one to two rows on 30- or 40-inch beds, but as with sudangrass, they can be sown on flat ground (with adequate drainage) using a grain drill. Some cowpea varieties are indeterminate, and the vines continue growing throughout the growing season; this must be taken into consideration to prevent them from disrupting irrigation systems with their excessive growth. No fertilizer is typically used in growing cowpeas, but sudangrass is fertilized with nitrogen and sometimes other nutrients. Emergence is rapid, about 4 days under warm summer conditions in the desert. Under most conditions, irrigation is not required for at least 1 week after emergence.

Cowpea cover crops are sometimes attacked by whiteflies or cowpea aphid that can reduce growth. Iron Clay cowpea and Trudan 8 sudangrass are cultivars with root-knot nematode resistance that are commonly used in this area.

Coastal Region

Richard Smith and Oleg Daugovish

Background

The coastal region has a moderate climate due to its proximity to the ocean. Typical daily maximum summer temperatures range from 70° to 80°F, and these moderate temperatures can extend as far as 40 to 50 miles inland, depending on the extent of cooling winds from the ocean. Vegetable production regions include the Salinas Valley, the Santa Maria Valley, the Oxnard Plain, and smaller valleys along the coast. Because of the moderate temperatures the region has a distinct agriculture that is dominated by cool-season vegetable production, with considerable warm-season vegetable production in southern California. Land values are typically high, which necessitates multiple crop rotations per year to cover costs. Cash crops are grown nearly year-round, but the majority of the planting occurs from December to August, with harvest from March to November. Due to the intensity of the agriculture it is difficult to find opportunities to fit cover crops into the rotations because cover crops may occupy land during the production season or cause delays in planting cash crops. As a result of these difficulties, typically only 5 to 7% of the vegetable acreage is cover cropped each year, based on a survey of local seed companies' sales. The percentage of land that is cover cropped in organic production is higher than that in conventional production.

Cover Crop Production

Cover crops are typically drilled or broadcast at common seeding rates (see chapter 2) and sprinkle-irrigated to initiate germination and growth. Varieties and planting dates are based on the purpose of the cover crop, disease concerns, and the impact the cover crop has on the vegetable planting schedule. When cover crops are mature they are incorporated into the soil in a variety of ways, typically by flail mowing and disking. Conservation tillage is not practiced in this region. Cover crop production is divided into winter, fall, and summer seasons.

Winter production

Winter is the most common season for cover crop production in this region. Planting dates extend from October to December, and cover crops are grown until February or March. Land that is cover cropped in the winter is typically not scheduled to be planted to a cash crop until later in the spring, so late-spring rains do not delay incorporation of the cover crop and disrupt the vegetable planting schedule. There is more impetus to cover crop certain problematic soils, especially if they are located in areas with cheaper land rents. *Sclerotinia minor* is the key soilborne disease of lettuce in this area, and lettuce growers do not plant cover crop varieties that are susceptible to this disease (i.e., peas, vetches, and phacelia). As a result, the dominant cover crops in this area are grasses such as cereal rye, barley, and oat. Organic growers with lettuce as a dominant crop in their rotations also tend to grow grass cover crops. However, organic growers who have diverse operations are less concerned with soilborne diseases of lettuce. Thus they frequently use legume-cereal mixes. The types of mixes can vary widely in species makeup, but they typically include a late cereal variety (e.g., Cayuse oat) at the rate of 15 to 20% by weight, with the remaining mix comprised of legume species such as bell beans, vetches, and field peas.

Water quality concerns are growing in importance in the coastal region; the primary concern is with sediment and nutrient loss from winter storms. Winter cover crops provide excellent protection against sediment and nutrient loss from runoff and also reduce nitrate leaching. A key challenge for the coastal region is how to include more cover crop acreage into planting schedules. One concept that is being examined is planting low-growing grass cover crops such as Trios 102 winter triticale on the furrow bottoms to reduce runoff, sediment, and nutrient loss (see chapter 4). The cover crop is terminated prior to planting the first cash crop with the expectation that the low-residue cover crop will not impede subsequent cash crop planting operations.

Fall production

The fall can be a useful cover crop slot for vegetable growers in the coastal region. This is particularly true if cash crop production is completed by August. In this scenario the cover crop is planted immediately after the cash crop and grown for 60 days. The cover crop is typically incorporated into the soil by mid-October to allow sufficient time for the ground to be worked and prepared for the winter fallow period. Typical short-term cover crops such as barley, mustards, and buckwheat are grown. Acreage of this planting slot has increased in recent years due to the rise in popularity of mustards planted as a short-term rotation with lettuce.

Summer production

Summer cover crops are little used in the coastal region because of high land rents. However, sudangrass is occasionally planted, and in organic operations, a mix of sudangrass and vetch is sometimes grown before planting overwintered vegetables.

12 *Laura Tourte and Karen Klonsky*

Economics

Economic Considerations

Farming operations include cover crops in vegetable rotations to achieve production, regulatory, and environmental goals. Production goals include building and maintaining soil fertility and quality; regulatory goals include compliance with relevant federal, state, and local legislation; and environmental goals include soil and water resource protection.

The economic impact of cover crops on vegetable production is difficult to accurately assess. Cover crops are typically not harvested and sold and therefore do not directly generate income. They do have direct establishment and management costs, as well as opportunity costs, which are discussed below. In some cases these costs can increase vegetable production costs beyond the cost of managing the cover crop. However, short- and long-term benefits may help reduce production costs for subsequent income-generating crops.

Benefits of Cover Crop Use

Legume cover crops provide nitrogen for crop production by cycling residual soil nitrogen or by fixing atmospheric nitrogen. The nitrogen provided by cover crops usually allows growers to reduce some nitrogen fertilizer use in the subsequent cash crop, thereby reducing the cost of production. However, the nitrogen savings associated with cover crop use depends on the type of cover crop grown, how it is managed, its nitrogen content and rate of release and availability, and the nitrogen needs of the subsequent cash crop.

Cover crops can have long-term impacts on soil fertility by contributing to the soil-building process. By adding organic matter to the soil, they improve soil structure, tilth, and overall quality over the long term. However, these long-term benefits are difficult to measure and value.

Growers also use cover crops as a conservation management practice to provide ground cover and improve water infiltration. Cover crops can capture residual soil nitrogen and reduce leaching of nutrients from the soil profile. They can be especially helpful in slowing surface water runoff from fields as well as in minimizing soil erosion on sloped or hilly land in areas with significant rain events during the winter months. In areas with high rainfall, some growers report a savings in labor and equipment use costs for farm and ditch cleanup after significant storm events (Tourte et al. 2003). When planted and maintained as a conservation practice, cover crops may ultimately contribute to the maintenance and protection of on-farm and downstream water quality, as well as to the protection of farmland and investments in property. Preventing or minimizing negative impacts to water quality and property damage may reduce conflicts with neighbors and exposure to legal and regulatory actions by state and regional government agencies.

Cover crops can also have beneficial impacts on pest management. They can help suppress soilborne diseases, reduce weed infestations, and provide habitat for beneficial arthropods. Assigning cost reductions to these effects is a difficult task. It must be borne in mind that cover crops can, in some cases, aggravate pest management problems.

Direct Costs

Costs of producing annually planted cover crops can vary but generally include land preparation (chiseling, disking), seeding (broadcast, drill), and sometimes irrigation to support germination and ensure good stand establishment, most notably in years with unusually dry weather. After a period of vegetative growth, cover crops are often mowed and disked to incorporate them into the soil and prepare for planting the subsequent cash crop. Sample cash costs for selected cover crops per seeded acre, as well as establishment and management cash costs per acre for

Table 12.1. Cash cost of seed per acre for selected cover crops, 2008

Cover crop	Seeding rate range (lb/seeded ac)	Price range* ($/lb)	Average cost† ($/seeded ac)
barley	90–100	0.32–0.50	39
bell beans (fava)	125–175	0.54–0.66	90
buckwheat	20–30	0.69–0.79	19
cereal rye, Merced	80–90	0.48–0.57	45
cowpea	40–60	1.25–1.44	68
mustard, Pacific Gold	7–10	2.99–3.35	27
mustard, Ida Gold	10–15	2.99–3.35	40
oat, Cayuse	80–90	0.39–0.54	40
pea, various	70–100	0.49–0.84	57
sorghum-sudangrass	25–30	0.65–0.75	19
triticale	100–120	0.41–0.69	61
vetch, common	55–75	0.95–1.04	65
vetch, Lana	40–60	1.30–1.84	79
vetch, purple	40–60	0.89–1.20	53
wheat	100–120	0.45–0.57	56

Notes:

*Depends on supplier and quantity purchased.
†Average seeding rate/acre × average price/pound.

Table 12.2. Sample cash costs per acre for an annually sown grass-legume cover crop blend, 2008

Operation	Labor		Material cost* ($/ac)	Total cost ($/ac)
	hr/ac	cost/ac ($)		
Land prep: chisel 1×	0.13	3	3	6
Land prep: disk 1×	0.14	3	4	7
Plant: drill seed	0.2	4	83	87
Irrigate: sprinkler	1.6	29	39	68
Mow: flail 1×	0.52	11	9	20
Incorp. plant materials: disk 2×	0.29	6	7	13
Operation costs		56	145	201
Interest on operating capital, 6.75%				2
Cash costs per acre				203

Note:

*Material cost includes seed, water, fuel, lube, and repairs.

an annual grass-legume cover crop mix, are shown in tables 12.1 and 12.2. Establishment and management costs are representative cash costs for equipment use, material inputs (seed, water, fuel, lube, and repairs), and labor.

Additional costs are sometimes associated with growing cover crops in vegetable systems. Besides the irrigation cost mentioned above, growers may continue to use standard fertilization practices after growing a cover crop, even one planted for its nitrogen contribution, to ensure top performance in the yield and quality of a vegetable crop. This is because of variability in release of nitrogen from a cover crop and availability of nutrients for the subsequent cash crop. The grower's goal in both cases is to minimize production risks, even though such risk management strategies may result in additional cash input costs.

Opportunity Costs

Growers do not experience any loss in farm revenue when cover crops are planted as a substitute to fallow land during idle production months. Foregone income can be assumed, however, when revenue-generating land is taken out of production to add a cover crop to the rotation schedule. For example, intensive vegetable operations in the coastal region often produce 2 to 2.5 cash crops per acre per year. In vegetable systems where cover crops are planted, the number of cash crops is often reduced to 1.5 to 2 crops per acre per year, resulting in a loss of revenue.

Cover crops also present management challenges in vegetable farming systems. For example, when planted as a substitute for fallow land, cover crops can delay bed preparation, spring planting, and early harvest of a vegetable crop, possibly leading to lower than expected "first-to-market" prices.

Cost Sharing

If cover crops are planted specifically as a conservation management practice, growers may be able to recoup partial cash costs through various cost-share programs of the USDA Natural Resources Conservation Service (NRCS) and local resource conservation districts (RCDs). NRCS programs include the Environmental Quality Incentives Program (EQIP), the EQIP Organic Program, the Conservation Reserve Program (CRP), the Agricultural Water Quality Enhancement Program (AWEP), and the Wildlife Habitat Incentive Program (WHIP). Cost-share amounts depend on the year, location, and size of the farm. Cover crops are also used by growers as a tool to protect and maintain on-farm and downstream water quality and to comply with California State Regional Water Quality Control Board (RWQCB) requirements.

References

Tourte, L., M. Buchanan, K. Klonsky, and D. Mountjoy. 2003. Central Coast conservation practices: Estimated costs and potential benefits for an annually planted cover crop. University of California, Davis, Agricultural and Resource Economics Web site, http://coststudies.ucdavis.edu/conservation_practices.

Tourte, L., R. Smith, K. Klonsky, and R. DeMoura. 2004. Sample costs to produce organic broccoli, Monterey and Santa Cruz Counties. University of California, Davis, Agricultural and Resource Economics Web site, http://coststudies.ucdavis.edu.

———. 2009. Sample costs to produce organic leaf lettuce, Santa Cruz and Monterey Counties. University of California, Davis, Agricultural and Resource Economics Web site, http://coststudies.ucdavis.edu.

Measurement Conversion Table

U.S. Customary	Conversion factor for U.S. Customary to Metric	Conversion factor for Metric to U.S. Customary	Metric
inch (in)	2.54	0.394	centimeter (cm)
foot (ft)	0.3048	3.28	meter (m)
yard (yd)	0.914	1.09	meter (m)
acre (ac)	0.4047	2.47	hectare (ha)
ounce (oz)	28.35	0.035	gram (g)
pound (lb)	0.454	2.205	kilogram (kg)
ton (T)	0.907	1.1	metric ton (t)
pound per acre (lb/ac)	1.12	0.89	kilogram per hectare (kg/ha)
ton per acre (T/ac)	2.24	0.446	metric ton per hectare (t/ha)
miles per hour (mph)	1.61	0.62	kilometers per hour (km/h)
Fahrenheit (°F)	°C = (°F – 32) ÷ 1.8	°F = (°C x 1.8) + 32	Celsius (°C)

Index

Page numbers followed by *f* refer to figures; page numbers followed by *t* refer to tables.